T0296123

CAMBRIDGE COMPARATIVE PHYSIOLOGY

GENERAL EDITORS:

J. BARCROFT, C.B.E., M.A.
Fellow of King's College and Professor of
Physiology in the University of Cambridge

and

J. T. SAUNDERS, M.A.
Fellow of Christ's College and Demonstrator
of Animal Morphology in the University of
Cambridge

CILIARY MOVEMENT

CILIARY MOVEMENT

J. GRAY, M.A.

Fellow of King's College
and Lecturer in Experimental Zoology in
the University of Cambridge

CAMBRIDGE

AT THE UNIVERSITY PRESS

1928

CAMBRIDGE
UNIVERSITY PRESS

University Printing House, Cambridge CB2 8BS, United Kingdom

Cambridge University Press is part of the University of Cambridge.

It furthers the University's mission by disseminating knowledge in the pursuit of education, learning and research at the highest international levels of excellence.

www.cambridge.org
Information on this title: www.cambridge.org/9781107502284

© Cambridge University Press 1928

First published 1928
First paperback edition 2015

A catalogue record for this publication is available from the British Library

ISBN 978-1-107-50228-4 Paperback

CONTENTS

PREFACE

Although ciliary movement was first observed by van Leeuvenhoek about the middle of the seventeenth century, it was not until 1834 that Purkinge and Valentin discovered the ciliated epithelium on the oviductal walls of the vertebrates. From that year, cilia may be said to have become the subject of serious physiological study.

In some cases the movements of an animal's body are due to the activity of muscles, but in others corresponding movements are effected by cilia. Since the mechanical powers of a cilium are slight in comparison to those of a muscle fibre, it is not surprising to find that most movements of large or active vertebrates are executed by muscles. In such organisms, cilia are restricted to limited regions of the body, and their functions, although important, are not spectacular. On the other hand, in very small animals or in invertebrates where the velocity of movement is very low, cilia and not muscle fibres often play the dominant rôle as organs of contraction and locomotion. The importance of ciliary movement in the life of many invertebrate animals has now become abundantly clear, and in most cases the ciliated surfaces are co-ordinated in a way quite unknown in higher animals. A comparative account of such ciliary mechanisms is urgently required, but in the present volume I have adopted a physiological rather than a morphological point of view.

Since muscular units are of overwhelming importance for the movement of vertebrate animals it is natural that physiological text-books should give only a brief account of the properties of ciliated epithelium. Those interested in the physiology of the lower animals may feel, however, that ciliary movement deserves more adequate analysis than it has yet received, and since so few facts have been established, it seemed desirable to set forth our knowledge in such a form as might indicate the possibilities of further research.

Any real conception of ciliary movement must eventually rest on a proper understanding of the hydrodynamical problems which are involved. At present, little is known of the forces which surround very small elongated structures when they are moving through water at very low speeds. In a few places I have tried to indicate the value of such knowledge and in doing so I have run considerable risk of error. Should such errors attract attention and thereby lead to a better understanding of the facts, one of the main purposes of the book will have been fulfilled. To even a casual reader it will be obvious that I have not restricted myself to a statement of established facts, but have expressed, too freely perhaps, hypotheses which are not always capable of experimental test.

To all those who have read and criticised the manuscript I owe my sincere thanks. I also have to acknowledge permission to reproduce many of the diagrams.

J. GRAY

King's College
  Cambridge
27 *October* 1927

Chapter I

INTRODUCTION

Ciliary movement may be defined as the work which a cell does by means of permanent but movable structures located at its surface. Such structures are found in almost all groups of the animal kingdom[1], and they play an important rôle in the organisation of the body.

It is only by means of cilia that a cell can perform mechanical work without itself undergoing marked changes in form. The physiological significance of this property is readily seen in the classical example of ciliated epithelium found on the oesophagus and palate of the frog (Fig. 1).

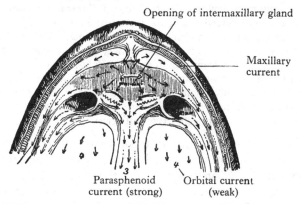

Opening of intermaxillary gland

Maxillary current

Parasphenoid current (strong)

Orbital current (weak)

Fig. 1. Ciliary currents on the roof of the frog's mouth. (After Merton.)

The function of this tissue, apparently, is to maintain the mouth in a reasonably aseptic state and to carry small particles of food or foreign objects into the alimentary canal. By the activity of the cilia, a thin sheet of water and mucus is kept moving parallel to the surface of the tissue without any change in the form or tension of the epithelium itself: it is obvious that muscular effort could only perform such a function with difficulty. The maintenance of a flow of water as a thin film over the surface of an organ is, in

[1] In all groups except Nematodes and typical Arthropods.

fact, a function peculiar to cilia, although it is by no means their sole use. A detailed account of the more complicated ciliary arrangements which are found in invertebrate animals is given in Chapter VIII, but it is interesting to remember that in many protozoa, and in many marine larvae, cilia are the only motile structures found in the body and it is by their means that the animals propel themselves through the water and obtain their food.

The simplest type of vibratile organ is the *flagellum* which is characteristic of flagellate Protozoa and of sponges. Each flagellum is a separate vibratile unit with its own blepharoplast, and it is usually of considerable length (see Fig. 2). As a rule one cell or individual protozoon does not possess more than four flagella. When a cell possesses numerous vibratile units these are usually much shorter than typical flagella and they are more commonly known as *cilia*. Cilia vary from $0 \cdot 1 - 0 \cdot 3 \mu$ in diameter and are often approximately 15μ in length. Bütschli (1889) estimated that a single *Paramecium* possessed 2500 cilia, whereas in *Balantidium elongatum* the number may be as high as 10,000. Most of the larger vibratile organs, usually referred to as cilia, represent a number of primary units associated together in varying degrees of intimacy. When these compound structures are more or less conical in form they are known as *cirri* (Fig. 5), when arranged to form a plate-like structure the term *membranella* is frequently employed (Fig. 4). For the sake of convenience the term 'ciliary movement' may be used to cover the movement of all these vibratile structures, and the term 'cilium' to cover both simple cilia and the less complicated types of 'cirri.'

Whereas muscular movement can occur either in water or in air, ciliary movement is restricted to water. The freedom of movement of a ciliated organism is thus limited by the physical properties of its medium to a much greater extent than is that of any terrestrial animal, and its study involves problems in hydrodynamics which are both interesting and difficult.

When a terrestrial animal jumps a distance of one foot vertically upwards in air by its own activity, a large proportion of the energy expended is used in overcoming the force of gravity. If the same animal were to carry out this movement in water, the amount of work done against gravity would be very much less owing to the fact that its specific gravity is not much higher than that of water (the work done against gravity being seldom more than one-tenth of that done in air). On the other hand, in order to move through the water a considerable amount of energy has to be expended in doing work against the resistance of the water. The viscous resistance of water is undoubtedly the factor which

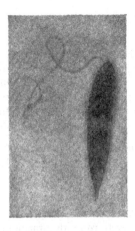

Fig. 2. Photograph of *Euglena*. Note the long flexible flagellum. (From Calkins.)

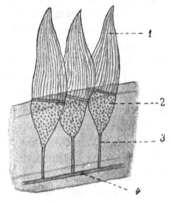

Fig. 4. Three membranellae from the adoral zone of *Stentor*. 1. Vibratile elements. 2. Basal lamellae. 3. Terminal fibres. 4. Basal fibre. (After Doflein.)

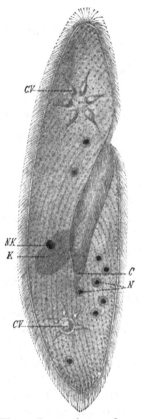

Fig. 3. *Paramecium caudatum* shewing the general distribution of the cilia. (After Doflein.)

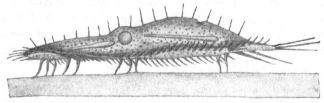

Fig. 5. *Stylonychia mytilus* in side view. Note the varied types of cilia present, particularly the large ventral cirri. (After Bütschli.)

limits the utility of cilia in all their functions (see Chapter IV), although at the same time it is the factor which makes ciliary locomotion possible. Up to a certain point an adequate analogy is provided by an aeroplane; the power of moving through the air by the activity of the engines depends on the fact that the air offers a resistance to the force exerted by the propeller, and at the same time it is the resistance of the air which limits the maximum speed of horizontal flight with a given engine.

When a cilium strikes the water the whole organism tends to move in a direction opposite to that of the effective beat, whereas the water tends to move in the direction of the beat. The nett result obviously depends upon the force required to set the organism in motion. If the animal is unable to move, or is very large, the water is driven over its surface and the animal remains stationary; if the animal is small and is not attached to any fixed object then the pressure exerted by the moving cilium tends to set the organism in motion at an initial velocity inversely proportional to its mass. This is the same principle by which a paddle wheel can drive a steamer through the water or drive water past a steamer which is fixed to the shore. Biologically, however, the principle has an important corollary. When a film of water is kept moving over the surface of a ciliated organ it is often of respiratory significance and the cilia beat without cessation during the whole life of the animal; so far as is known, no animal is able to interfere with the beat of the cilia which drive such surface currents[1], nor would such a power be of any apparent physiological value. On the other hand, when the conditions of ciliary movement are such that the whole organism is thereby driven through the water, we find that the animal is able to modify the movement of the cilia in such a way as to change the direction of its motion or cease moving altogether. The control of cilia will be considered in greater detail, but it appears to afford an interesting example of the way in which organisms have evolved a system of control in response to a simple mechanical principle. There can be little doubt that all cilia are fundamentally automatic in their movement (see p. 123) and that the power possessed by organisms to inhibit their locomotory cilia is of an extraneous nature.

[1] McDonald, Leisure, and Lenneman (1927) have recently reported, however, that the pharyngeal cilia of the frog beat more rapidly when their sympathetic nerve supply is stimulated, and less rapidly during stimulation of the parasympathetic system.

The small size of even the largest flagella and cilia not only limits the methods available for their study but raises difficult problems of dimensions. Cilia, which appear under the microscope to be moving with extreme rapidity, are actually moving with a very low velocity. It is important to realise that all ciliary movement is essentially slow (see Chapter II), although it may produce results superficially similar to those of more rapid muscular movement. In terms of their own dimensions a flagellated spermatozoon is probably moving at approximately the same relative speed as an average fish, but the absolute velocity of the latter is very much greater.

Table I.

(See *Tabulae Biologicae*, vol. IV.)

Organism	Absolute speed of movement (per sec.)	Relative speed =multiple of organisms' length in 1 sec.
Bacillus subtilis	10μ–15μ	5
Spirillum volutans	110μ	8·5
Euglena spec.	115–235μ	3–5
Paramecium	1·3 mm.	6
Spirostomum	600μ	0·6
Halteria saltans	2·3 mm.	77
Amoeba	$0·5\mu$–$2·5\mu$	0·02
Alectrion (ciliary movement)	2·5 mm.	0·01
Fish	50–200 cm.	1–4
Spermatozoa (fish)	33–180μ	2–8
Man (walking)	150 cm.	0·75

There is probably no great difference in mechanical principle in the modes of progression of a spermatozoa and a fish; in each case the lateral movements of the body follow one another in such a way as to impart a backward thrust on the water, the reaction of which enables the animal to move ahead; but, when we attempt to analyse in any detail the relative efficiency of the two mechanisms, we find important differences. In each case most of the energy expended during horizontal movement is used up in doing work against the frictional resistance of the water, but the laws which determine the relationship between the velocity of movement and the frictional resistance of a large and rapidly moving body cannot be applied to small bodies moving at very low speeds. At present our knowledge of the work done by cilia is extremely small, and until further experimental evidence is available many of the more fundamental problems must remain obscure (see Chapter IV). For slow continuous movements involving only a low horse-power cilia are of great value, but where the rate of doing work is high and where it is important to move suddenly at high speed the advantage to be derived from a muscular mechanism is very marked except in the case of very small animals (see Chapter IV).

The following table indicates to some extent the types of organs in which cilia play an essential rôle. It will be noticed that the

Table II.

Type of organ	Typical example	Ciliary function
Epidermis	(a) Most invertebrate larvae, Ctenophores	Locomotion
	(b) Turbellaria	Locomotion
Alimentary canal	(a) Frog's oesophagus	Maintains superficial current
	(b) Intestine and liver of molluscs	Propulsion of fluid through narrow tubes, the walls of which are ciliated
Excretory system	Nephrostomes and nephridia of most invertebrates	Propulsion of fluid through narrow tubules
Reproductive system	(a) Vasa efferentia in vertebrates	Propulsion of fluid through narrow tubules
	(b) Fallopian tubes in mammals	Moves ova into uterus
External appendages	(a) Disc of Rotifers	Locomotion and nutrition
	(b) Tentacles of Polychaets, Polyzoa, etc.	Maintains superficial nutritive current
Sensory	(a) Ciliated tracts of Ctenophores	? Conduction of stimulus
	(b) Eyes of Molluscs Otocysts of Pteropods	?? Sensory receptors

propulsion of water may take place either over a flat surface or over the inner walls of a tube (see Chapter III).

The morphological structure of the vibratile elements.

One theory after another has been advanced to elucidate the essential mechanism of a cilium, but no convincing correlation has yet been made between the structure as seen under the microscope and what is known of the changes occurring in the living cell. When we remember that really nothing is known of the

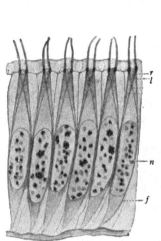

A. Six latero-frontal cells of the gill of *Mytilus edulis. n,* nucleus; *f,* common fibril; *r, l,* ciliary rootlets. (After Bhatia.)

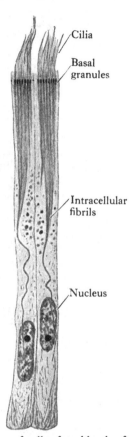

Fig. 6. *A* and *B.*

B. Side view of cells of typhlosole of *Anodon.* Note well-defined intracellular system of fibres. (After Gurwitsch.)

machinery whereby a muscle develops a tension when it is stimulated, it is not surprising that the ciliary mechanism also remains obscure. In one respect, however, ciliated cells invite more sweeping speculation than is usually expended on muscular units for, in contradistinction to the latter, the morphological structure of the

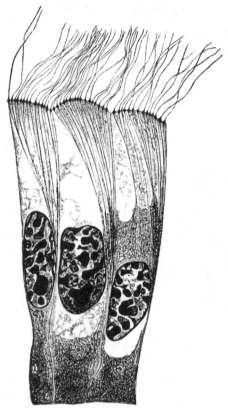

C. Surface view of cells in hepatic duct of *Helix*. (After Heidenhain.)

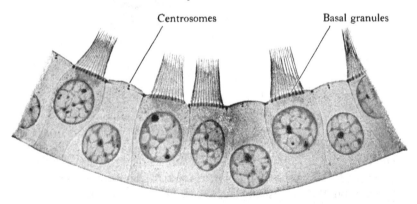

Centrosomes   Basal granules

D. Ciliated and non-ciliated cells of epididymis of a puppy. Note absence of intracellular fibres: also the peripheral 'centrosomes' in the non-ciliated cells. (After von Lenhossék.)

Fig. 6. *C* and *D*.

cells is extraordinarily constant no matter what is the organ or the animal in which they are found (Fig. 6, *A–D*).

For the detailed morphology of the ciliary rootlets and basal granules reference should be made to suitable monographs (Erhard, 1910; Prénant, 1914; Saguchi, 1917). In its simplest form the vibratile element shews no optical structure even under polarised light. When viewed by transmitted light or on a dark background both cilia and flagella are optically homogeneous. When viewed under polarised light an apparent doubly refracting property is observed (Engelmann, 1898), but there can be little doubt that the phenomenon is due to diffraction at the surface of the flagellum or cilium and not to its internal structure (Mackinnon and Vlès, 1908).

In a limited number of cases large flagella appear to shew a definite structure in 'preserved' material. According to Calkins (1926) "the flagellum is made up of two definite elements, an axial highly vibratile filament which is formed as an outgrowth from the basal body or blepharoplast, and an enveloping elastic sheath which is formed from the protoplasmic substance of the cortex. In some cases the sheath is circular in cross-section, in others ellipsoidal, while the contractile thread, which is usually firmly attached to the sheath, may run in a straight line the entire length of the sheath or may follow a spiral course. In the majority of flagellates the sheath undulates and vibrates in unison with the contractile axial thread, but in a few types such as *Peranema trichophora* or certain species of *Astasia* the sheath remains passive while the axial thread extends freely beyond the limits of the sheath where its activity

Fig. 7. *a.* Flagellum of *Trachelomonas* shewing axial filament and sheath. *b.* Transverse section of the flagellum. (After Doflein.)

in the surrounding medium results in a steady progressive movement of the cell" (p. 134). Wenyon (1926) describes the flagellum as follows: "A flagellum, as pointed out by Alexeieff (1911), consists of an axial filament (axoneme) and a thin sheath of cytoplasm. The axoneme itself takes its origin in a minute granule, the *blepharoplast*, which is situated in the cytoplasm, and sometimes upon the surface of the nuclear membrane. The axoneme passes to the surface of the body and there acquiring a thin sheath of cytoplasm, becomes the flagellum" (p. 31).

It is very doubtful whether any of these structures can be seen in the living flagellum, although the development of the tail of many spermatozoa undoubtedly supports the view that in the early stages two elements are present, viz. an axial filament and a 'protoplasmic' sheath. It is by no means clear, however, that the movements of a typical flagellum demand the presence of both structures (see Chapter II).

As already mentioned, nearly all the larger cilia occurring in nature can be resolved into a number of smaller units each of which possesses the power of movement as long as it is attached to the cell. According to Taylor (1920) the cirri of *Euplotes patella* are to be regarded as a series of vibratile filaments embedded in a viscous or sticky matrix. On the other hand Carter (1924) shewed that the *laterofrontal* cilia on the gills of *Mytilus* are normally composed of a series of triangular plates in close contact with one another, but no matrix appears to be present. The edges of the plates are parallel to the plane of beat, and the surfaces of the plates are at right angles to this; the plates nearer to the end of the effective stroke are shorter than those further away. After displacement from their normal position individual plates readily regain their normal position and beat in unison. A comparable structure occurs in the 'undulating membrane' (*membranella*) of *Blepharisma* (Chambers and Dawson, 1925). On puncture with a needle (Fig. 9) this membrane at once splits along a line running through the puncture and part of the membrane breaks into a series of very fine cilia which beat out of unison. As soon as the needle is removed from the membrane the row of isolated cilia quickly reforms an apparently homogeneous membrane, the elements of which beat synchronously.

Fig. 8. Laterofrontal cilium of *Mytilus* distorted by a needle (*p*). The whole cilium is composed of a series of triangular platelets similar to that seen to the left of the needle. (After Carter.)

As a rule, however, even quite large cirri, such as are found in heterotrichous ciliates, appear homogeneous in life, and their compound nature is only shewn by the number of basal granules concerned and possibly by their mode of origin (cf. also

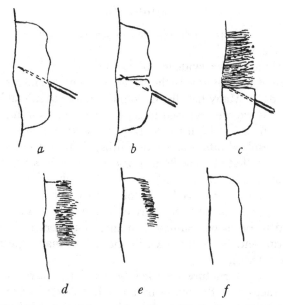

Fig. 9. Microdissection of the membranella of *Blepharisma undulans.* When the membranella is held by the needle, as in *a*, it splits at the point of contact, and the region beyond the needle breaks into a series of very fine cilia which beat out of unison. On removing the needle the cilia rapidly unite again (*d–f*) to form an apparently homogeneous membrane. (After Chambers and Dawson.)

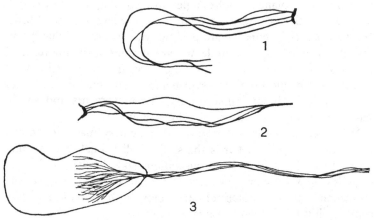

Fig. 10. 1 and 2. Uncoiled fibrils of the flagellum of *Euglena* after treatment with osmic acid. In 3 note the intracellular rootlets of the fibrils forming the flagellum. (After Dellinger.)

Ctenophores, where each platelet arises as a number of small vibratile elements).

The whole of the evidence points to the conclusion that the fundamental motive unit is the flagellum whose diameter may not exceed 0·05 μ. Many flagella, of course, are of considerably larger diameter. If we are prepared to accept evidence from histological methods, there can be little doubt that the typical flagellum can itself be resolved into a series of fibrils. Fig. 10 from Dellinger (1909) shews that the flagellum of *Euglena* (which is undoubtedly a single morphological unit in the sense that only one basal granule is present) can be resolved into four fibres spirally twisted over each other. This fibrous structure of a simple vibratile unit is not restricted to Protozoa; similar structures have been observed in other forms and are well known in the tails of many spermatozoa (Ballowitz, 1890).

How far the structure revealed by fixation is a reliable guide to the structure of the living unit requires careful scrutiny, but as far as it goes it indicates that the problem of ciliary movement can be resolved into finding out how the cilia can develop a tension in their fibrous elements, just as is the case in a muscle cell. The difference between a cilium and a muscle cell lies in the fact that in the former case a disturbance takes place in the fibrils in a plane at right angles to their length, whereas in a muscle fibre the disturbances occur along the length of the fibrils. In other words in cilia we are dealing with transverse changes in the fibrils, whereas in muscle we are dealing with longitudinal changes. It is interesting to note that in the flagellum or tentacle of *Noctiluca* both transverse and longitudinal disturbances occur, since it shews transverse flagellar movements and at the same time contracts and expands longitudinally.

Whatever be the essential parts of the ciliary machine we can be tolerably certain that it is the same in all animals. As already mentioned, the acidophil fibrous cilium and the basophil basal granules shew little or no variety of form or position, whereas the corresponding morphological structures in muscle fibres vary considerably in different groups of animals.

The nature of the ciliary mechanism has long been the subject of morphological discussion, and no serious attempt will be made to gather together the numerous theories which have been put forward. Adequate reviews from the historical and other points of view have been made by Pütter (1903), Erhard (1910), and Prénant (1914). There are, however, a limited number of facts which are of value, and which must eventually be woven together into a complete theory. In the first place, it is certain that the motivity of the cilium is independent of the cell structure as a whole. Peter (1899) was the first to shew that when a ciliated protozoon is crushed into quite small fragments the cilia continue to beat as long as they are in organic connection with a fragment of cytoplasm. Erhard (1910) and Gray (1926) have shewn that profound changes can occur in the proximal cytoplasm and in the nucleus without entailing a cessation of ciliary movement. According to Engelmann (1898) enucleated spermatozoa exhibit active movement. There is, therefore, general agreement that the whole of the essential ciliary mechanism lies at the distal end of the cell. Beyond this point there is marked divergence of opinion. Peter

Fig. 11. Excised strip of lateral epithelium of *Mytilus* (diagrammatic). Note that the cilia are active as long as they are in organic communication with the cells.

observed that cilia completely isolated from the distal protoplasm are motionless, and since basal granules are always present in this region it may not unreasonably be suspected that they are the kinetic centres of the system. Engelmann (1898) stated that if the tail of the frog's spermatozoon be cut off from the 'middle piece' all movement ceases, whereas if the cut is made between the nucleus and the 'middle piece' active movement continues. A similar state of affairs can be observed in the *lateral* epithelium of the gills of *Mytilus* (see Fig. 11) where the cilia can be stripped away from the cells, leaving the basal granules behind; such cilia are always motionless. (See also v. Rényi, 1926.)

The views of Peter received considerable support from the so-called 'Henneguy-Lenhossék theory' of the origin of the basal granules. According to these authors the basal granule of a cilium is homologous and sometimes identical with the centrosome of cell division. Just as the centrosome was regarded as the kinetic centre for nuclear division, so in its rôle as basal granule it was the kinetic centre for ciliary movement. It is impossible to discuss in any detail the cytological evidence in support of these conclusions, although it is in some cases remarkably strong. The latest adherents to Henneguy's view are Jordan and Helvestine (1922), who claim that the ciliated cells in the epididymis of the rat divide amitotically because the division centres of the cells are functioning as basal granules[1].

On the other hand, these views have not been unchallenged. Erhard (1910) claims that the basal granules can be destroyed by heat without affecting the motivity of the cilia; Meves (1899) denied that the middle piece is necessary for the movement of the spermatozoon of the guinea-pig. Again, some actively ciliated cells apparently lack basal granules, although they are present in the immovable 'Stiftchenzellen' of the molluscan eye (Maier, 1903) and similar structures (Hesse, 1900).

The small size of the units concerned makes it difficult to subject the distal end of the cell to critical microdissection. When the living cells are examined microscopically the region occupied by the basal granule is usually characterised by an area of high refractive index but no definite structure is visible. By appropriate means (Gray, 1926) the protoplasm and the nucleus of the cells on the gills of *Mytilus* can be made to absorb water, and the basal granules can then be identified as a well-defined ridge of highly refractive material lying at the base of the cilia: we may therefore conclude that as far as the basal granules are concerned the essential structure of the cell is not materially altered by histological technique.

The cilia of the Protozoa appear to arise as protrusions of fine hyaline processes from the distal surface of the cell, and from a very early stage these processes shew active movement (Prowazek, 1902). In certain cases the mode of formation and movement of newly-formed cilia recall the origin and behaviour of pseudopodia (axopodia), and some authors have regarded these cases as of phylogenetic significance. Correspondingly, the pseudopodia of *Mastigella vitrea* appear to shew features in common with true cilia (Goldschmidt, 1907). In Metazoa the origin of cilia has not yet been satisfactorily demonstrated in the living cell (see Gurwitsch, 1899; Joseph, 1902).

[1] Kindred (1926) states that, although the basal granules in the cells of a frog's oesophageal epithelium are derived from centrosomes, both these structures are present in the mature cell.

In the following chapters an attempt is made to investigate the ciliary mechanism by an examination of the living cell. The facts are clearly insufficient to justify any correlation with morphological structure but they may at some distant date suggest a clue by which this will be done.

References

Alexeieff, A. (1911). C.R. Soc. Biol. 71, *379*.

Ballowitz, E. (1890). Pflüger's Archiv, 46, *433*.

Bütschli, O. (1889). Bronn's Tierreich. Bd. 1, *2035*.

Calkins, G. N. (1926). *Biology of the Protozoa*. London.

Carter, G. S. (1924). Proc. Roy. Soc. 96 B, *115*.

Chambers, R. and Dawson, J. A. (1925). Biol. Bull. 48, *240*.

Dellinger, O. P. (1909). Journ. Morph. 20, *171*.

Engelmann, T. (1875). Pflüger's Archiv, 11, *432*.

*—— (1898). *Dictionnaire de Physiologie*. Paris.

Erhard, H. (1910). Archiv f. Zellforsch. 4, *309*.

*—— (1922). Abderhalden's Handb. Biol. Arbeitsm. Abt. v, 2, 3, *213*.

Goldschmidt, R. (1907). Archiv f. Protist. Suppl. 1, *83*.

Gray, J. (1922). Proc. Roy. Soc. 93 B, *104*.

—— (1926). Brit. Journ. Exp. Biol. 3, *167*.

Gurwitsch, A. (1899). Arch. f. Mikr. Anat. 57, *184*.

Hertwig, R. (1877). Jen. Zeit. 11, *324*.

Hesse, R. (1900). Zeit. f. wiss. Zool. 68, *379*.

Kindred, J. E. (1926). Journ. of Morph. and Physiol. 43, *267*.

Jordan, H. E. and Helvestine, F. (1922). Anat. Record, 25, *7*.

Joseph, H. (1902). Arb. aus d. Zool. Inst. d. Univ. Wien, 14.

Mackinnon, D. L. and Vlès, F. (1908). Journ. Roy. Micro. Sci. *553*.

Maier, H. N. (1903). Archiv f. Protist. 2, *73*.

McDonald, J. F., Leisure, C. E. and Lenneman, E. E. (1927). Proc. Soc. Exp. Biol. and Med. 24, *968*.

Merton, H. (1923). Pflüger's Archiv, 198, *7*.

Meves, F. (1899). Arch. f. Mikr. Anat. 54, *381*.

Peter, K. (1899). Anat. Anz. 15, *271*.

*Prénant, A. (1914). Journ. de l'Anat. et Physiol. 49, *545* seq.

Prowazek, S. (1902). Arb. aus d. Zool. Inst. d. Univ. Wien, 14.

*Pütter, A. (1903). Erg. der Physiol. II, 2, *1*.

v. Rényi, G. (1926). Zeit. f. d. gesamte Anat. Abt. 1, 81, *692*.

Rhumbler, L. (1899). Zeit. Wiss. Zool. 46, *549*.

*Saguchi, S. (1917). Journ. Morph. 29, *217*.

Taylor, C. V. (1920). Univ. Calif. Publ. Zool. 19, *404*.

Wenyon, C. M. (1926). *Protozoology*, vol. 1. London.

* References of general interest.

Chapter II

THE MOVEMENT OF A CILIUM

The movement of an individual cilium.

When observed under the microscope most cilia appear to be moving with very high velocity, and it seems difficult to accept the statement that even the most rapid types seldom attain a velocity of twenty feet per hour. A simple calculation, however, rapidly disposes of any conception of a cilium as a rapidly moving unit. Let the length of a cilium be 10μ and let it oscillate through an amplitude of $180°$ twelve times every second; the total distance travelled in 1 sec. by the tip of the cilium is therefore

$$12\left[\pi\left(10\mu\right)\right] = 375\mu \text{ (approx.).}$$

According to Kraft (1890), the velocity of the effective stroke is five times that of the recovery stroke so that during the former phase the tip of the cilium will move with an approximate velocity of 1 mm. a second or 12 feet per hour. Without allowing for changes of velocity during the complete cycle, Bidder estimates the velocity of the tips of the flagella of sponges at 14 feet per hour. "We forget, as we look through a microscope...that though distance is magnified, time is not magnified" (Bidder (1923), p. 12). It must not be forgotten, however, that these slow linear velocities are associated with angular velocities of quite a high order; in the first example given, the cilium is moving with the angular velocity of a flywheel running at 360 revolutions a minute.

The difficulty of observing the beat of a healthy cilium under the microscope is not so much due to its apparent high linear velocity as to the short time occupied by one complete cycle, $e.g.\ \frac{1}{12}$ sec.; for the eye fails to distinguish clearly the series of events which occur in such a short space of time. Nearly all observations of the form of a cilium during the different phases of its beat are based on tissues whose activity has been artificially subdued by appropriate means to about one-tenth of the normal speed, so that each complete beat occupies approximately one second; the magnification used is such as will allow the eye to focus the cilium clearly throughout the whole cycle.

Since practically the whole of a cilium's energy, during horizontal movement, is used in doing work against the frictional resistance of the medium (see Chapter IV), a simple means of reducing the velocity of movement is provided by increasing the viscosity of the medium in which the cilia are moving. For this purpose, gum arabic or similar substances can be used, although reasonable care must be exercised to make the medium of uniform viscosity. Alternatively, the rate of movement can be reduced by lowering the temperature, or by raising the concentration of hydrogen-ions within the cells (see p. 80). Although we have no reason to believe that the form of the beat is fundamentally different at different speeds, it must be remembered that any marked change in the velocity of the beat may affect its form (see p. 28).

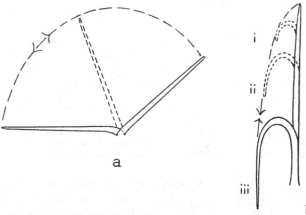

Fig. 12. The broken lines indicate the paths traced out by the tips of the cilia; the arrows indicate that, in these particular examples, the path during the effective stroke is approximately the same as that during the recovery stroke. *a*. Pendular movement; the cilium bends only at its base, *e.g.* in heterotrichous ciliates. *b*. Flexural movement; the cilium is straight when at rest, but during movement it bends along its whole length, *e.g.* latero-frontal cilia on the gills of *Mytilus*.

The form of moving cilia and flagella.

The simplest type of movement is the *pendular* movement (Fig. 12 *a*) characteristic of many of the larger compound cilia, and is frequently exemplified in the heterotrichous ciliates. In these cases the cilia vibrate backwards and forwards by flexure at the

G

2

base. The forward effective stroke is more rapid than the backward preparatory stroke but there is no obvious difference in the form of the cilium during the two phases; the tips of the cilia move through the same path during the two strokes. Regarding the cilium as an elastic structure, it is clear that the movement resembles that of a more or less rigid body attached to an elastic base.

A more common type of simple movement is that characteristic of the *latero-frontal* cilia of lamellibranch molluscs. In this case the cilium is perfectly straight when at rest (Fig. 12 *b*). Movement occurs by a flexure which begins at the tip and passes down to the base thereby bending the cilium into a hook-shaped structure. During recovery the process is reversed and the cilium straightens from the base to the tip.

Fig. 13. Combination of pendular and undulatory movements, *e.g.* in mammalian epididymis. (After Pütter.)

Fig. 14. *Trypanosoma balbianii* in motion; viewed along axis of movement. (After Perrin.)

Both these comparatively simple types of movement are usually restricted to vibratile organs which are of a comparatively sluggish habit and which are often of considerable size. The movement characteristic of the simplest vibratile element, viz. the flagellum, is of the undulatory type (Fig. 25). In this case a series of waves passes along the flagellum from its base to the tip or *vice versa.*

From these three main types there can be derived the form of movement of nearly every known vibratile organ. A combination of the pendular and undulatory movements is seen in the mammalian epididymis (Fig. 13) (Pütter, 1902). The motion of the frontal cilia of *Mytilus* (Fig. 17) represents a combination of the pendular and the hook-like movement, the details of which will be considered later.

Neither pendular nor undulatory movements necessarily occur in one and the same plane during all phases of the beat. The tip of a pendular cilium or of an undulatory flagellum may trace out an elongated ellipse when viewed from above (van Trigt, 1919). In the case of *Trypanosoma balbianii* the movement appears to be more complicated than in other forms (Perrin, 1906); it is illustrated in Fig. 14.

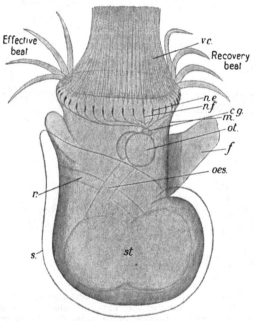

Fig. 15. Veliger larva (diagrammatic). *v.c.* velar cilia; *n.e.* nerve ending to ciliated cells; *n.f.* nerve fibrils; *c.g.* cerebral ganglion; *ot.* otocyst; *f.* foot; *m.* mouth; *oes.* oesophagus; *r.* rectum; *s.* shell; *st.* stomach. (After Carter.)

These modifications of vibratile movement have been known for many years (Valentin, 1842) but it is only comparatively recently that any detailed description has been given of the behaviour of those units (excluding flagella) which are known to perform a considerable amount of mechanical work.

The first of the more modern observations are those of Williams (1907). Unfortunately the parent organism is only described as "an unidentified but common larva of a protobranch mollusc" from Narragansett Bay. The large velar cilia (see Figs. 15 and 16), by which the larva moves, are arranged along the edges of an over-hanging groove, each cilium being somewhat curved with its concave face towards the side of the effective stroke (Fig. 16 a, A). This position can be readily observed whenever the animal comes

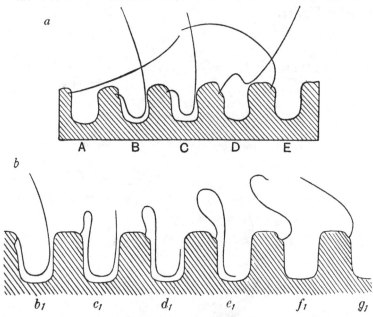

Fig. 16. a, A. Position of rest of velar cilium as described by Williams. B. Backward preparatory stroke; note that the cilium fits into the velar groove. C, D. opening phases of forward effective stroke; note that the cilium is crimped against the side of the groove. In D the cilium has just released itself from the groove and rapidly flies forward to position E. From E it swings back to the position of rest at A. The cilium is therefore more flexible during the backward preparatory stroke than during the forward effective stroke. [*Were the cilium equally flexible in the forward and backward strokes— forward stroke would occur as shewn in Fig. 16 b, b_1—g_1.*]

to rest, but the details of the beat can only be seen in larvae of reduced activity (*e.g.* when under a coverslip). In its resting position (Fig. 16 a, A) the cilium lies wholly outside the adjacent groove and the first phase of the beat is such that the base of the cilium moves backwards to fit quite closely into the groove (Fig. 16 a, B).

This is the end of the backward or preparatory movement. The forward stroke of the cilium now takes place and is divided into two distinct phases: (i) the base of the cilium begins to move upwards and forwards, whilst the overhanging lip of the groove prevents the movement of the cilium as a whole (Fig. 16 a, C and D); (ii) as the forward movement of the base continues the more distal part of the cilium eventually ceases to be crimped against the edge of the groove, and is suddenly released; it flies forward, with a very rapid effective stroke, far past its position of rest (Fig. 16 a, E). Finally the cilium swings back to its position of rest and the cycle is repeated.

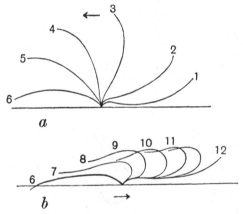

Fig. 17. a. Forward effective stroke of frontal cilium of *Mytilus*. Note the rigid form during the whole stroke. b. Backward preparatory stroke of frontal cilium of *Mytilus*. Note the flexible nature of the cilium. The flexure begins at the base and spreads to the tip. (After Gray.)

The essential features to note are (i) during its preparatory stroke, the cilium accommodates its form to that of the groove and is obviously very easily flexed, (ii) during the forward stroke it is not so easily flexed but offers considerable resistance to the edge of the groove. (Were the cilium equally flexible during both strokes the form of the effective stroke would be that shewn in Fig. 16b.)

The form of the beat of the frontal cilia on the gills of *Mytilus* was described by the author (Gray, 1922) and is shewn diagrammatically in Fig. 17. In this case the difference in flexibility during the two phases of the beat is very clear. During both the effective and recovery strokes the cilium is moving against the resistance of the water; but during the effective stroke the cilium behaves as a fairly rigid rod fixed at one end to the cell and its tip traces an arc of 180°; during the recovery stroke on the other hand the cilium moves backwards as a limp thread along which a backward stress is passing from base to tip. Both effective and recovery strokes take place in the same plane. These observations were confirmed by Carter (1924) for the large abfrontal cilia on the gills of *Mytilus*, who shewed that during the effective beat the whole cilium is rigid throughout its whole length: during the recovery

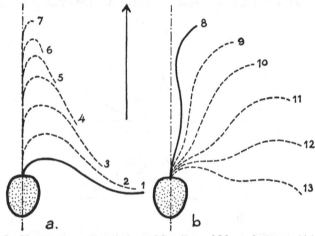

Fig. 18. Simplest type of movement of flagellum of *Monas* during rapid forward movement. *a*, 1–7. Successive stages in preparatory stroke. Note the flexure begins at the base and spreads to the tip. *b*, 8–13. Successive stages in the effective stroke. Note the rigidity of the cilium. The arrow indicates the direction of movement of the organism. (After Krijgsman.)

beat the cilium is limp and frequently slips below the micro-dissecting needle which is used as an obstruction. These properties remain for some time after the cilium is brought to rest for they can be observed by moving the cilium artificially through the two phases of its beat.

Quite recently Krijgsman (1925), using dark ground illumina-tion, has analysed the movements of the flagellum of *Monas*. This protozoon can move forward either slowly or rapidly; it can move backward, and it can move laterally. The simplest movements of the flagellum during forward movement are shewn in Fig. 18.

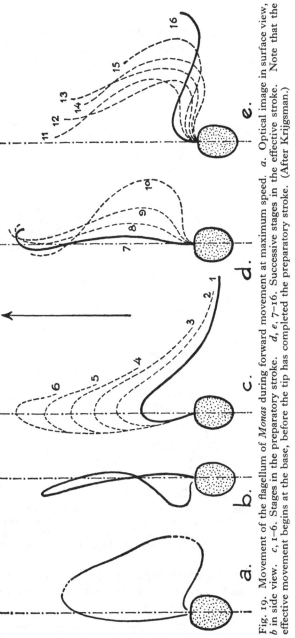

Fig. 19. Movement of the flagellum of *Monas* during forward movement at maximum speed. *a*. Optical image in surface view, *b* in side view. *c*, 1–6. Stages in the preparatory stroke. *d*, *e*, 7–16. Successive stages in the effective stroke. Note that the effective movement begins at the base, before the tip has completed the preparatory stroke. (After Krijgsman.)

It will be noted that the form of the flagellum during the two phases of the beat is essentially the same as that of the frontal cilia of *Mytilus*. The recovery stroke (Fig. 18 *a*) takes place by means of

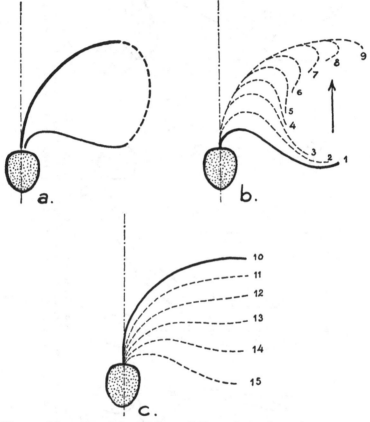

Fig. 20. Movement of the flagellum of *Monas* during forward movement at reduced speed. *a*. Optical image (surface view). *b*, 1–9. Preparatory stroke. *c*, 10–15. Effective stroke. Note the amplitude is less than in Fig. 19 and that the effective stroke takes place synchronously along the whole flagellum. (After Krijgsman.)

a bending movement which begins at the base of the flagellum and spreads to its tip. In the effective stroke the flagellum moves back as a more or less rigid structure (Fig. 18 *b*).

A more complicated type of rapid forward movement is shewn

in Fig. 19. In this case the tip of the flagellum appears to lag
behind the proximal portion during the effective stroke. It is of
interest to note that this phenomenon can be induced in the frontal
cilia of *Mytilus* by exposing them to an atmosphere containing
ammonia.

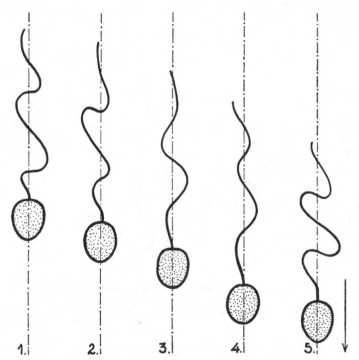

Fig. 21. Backward movement of *Monas*. Note that the pendular type of
movement is replaced by one of the undulatory type. (After Krijgsman.)

When *Monas* moves forward slowly, the form of the strokes
is essentially unchanged (Fig. 20) but the tip of the cilium instead
of passing through an arc of 90°, only moves through 45° or less.

When *Monas* moves backward, the flagellum is characterised
by the formation of waves which pass down from the cell to the
tip (Fig. 21); when moving to one side the flagellum is bent at
right-angles to the body and waves of distortion pass along the
flagellum from the base to the tip (Fig. 22).

One of the most interesting movements is seen in Fig. 23. In this case the whole flagellum swings through a small arc, but the distal portion alone exhibits small but numerous waves.

The restriction of movement to the tip of a flagellum is not an uncommon phenomenon in the Protozoa. A well known example is *Peranema* (Fig. 24).

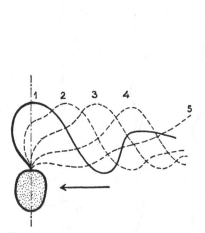

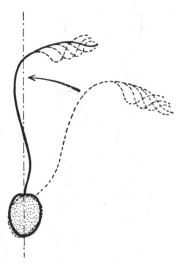

Fig. 22. Lateral movement of *Monas*. Note flexure of the flagellum at the base, and undulatory movements. (After Krijgsman.)

Fig. 23. Turning movement of *Monas*. Note that undulations of reduced amplitude are restricted to the tip of the flagellum. (After Krijgsman.)

The observations of van Trigt (1919) shew very clearly how the flagellum of *Spongilla* may also exhibit marked variations in the form of movement, and how these variations are associated with varying degrees of activity (Figs. 25, 26).

When moving with its normal rapid frequency, the flagellum of *Spongilla* exhibits a series of waves whose amplitude[1] and wave length are both small. These waves pass from the base to the tip of the flagellum with considerable velocity and a current of water is thereby driven forwards from the cell in a direction parallel to

---

[1] By 'amplitude' is meant the extent of the displacement of any point on the flagellum in a line at right-angles to the main axis of the moving flagellum. By 'wave length' is meant the length of flagellum occupied by one complete wave.

the long axis of flagellar movement (Fig. 25 A, *a*). Soon after isolation from the animal, however, the activity and frequency of vibration decrease, and at the same time the amplitude and the wave length of the individual waves increase (Fig. 25 A, *b* and *c*). Eventually the wave length and amplitude of the slowly moving flagellum become so large that the undulatory type of movement appears to pass into one of lateral displacement only, as in Fig. 25 A, *d*; finally the flagellum comes to rest in a more or less straight condition. There can be little doubt that during typical undulatory movement the flagellum forms part of a helix and that the diameter of the helix increases with decrease in the frequency of vibration. Van Trigt states that in some cases there is no doubt that the point of the flagellum during rapid movement travels in an elliptical orbit, but from his description one may infer that the axes of the ellipse are not increased uniformly when the 'amplitude' is increased. This point is of some theoretical importance and will be discussed later.

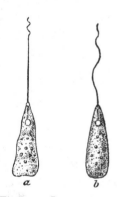

The undulatory type of movement as seen in a free-swimming organism is well illustrated by *Spirillum volutans* (Riechert, 1910; Fuhrmann, 1910). As in *Spongilla* rapidly moving flagella are characterised by short waves of small amplitude; when the speed of movement falls the amplitude and the wave length both increase. If the figures given by Fuhrmann (Fig. 27) are correct representations of the living organisms, they provide overwhelming evidence in favour of the view that the kinetic energy of the flagellum is generated along the whole length of the vibratile element and is not simply supplied to it at one end by mechanical movements going on in the cell. It will be noted that in Fig. 27, *b* and *c*, there is no apparent change

Fig. 24. *Peranema. a.* Slow forward movement with undulations restricted to the tip of the flagellum. *b.* Rapid forward movement with undulations along the whole length of the flagellum. (After Verworn.)

in the amplitudes of the waves seen in different regions of the flagellum, and this cannot occur if the energy of the waves is all provided at their source (see p. 42). Riechert's observations are of

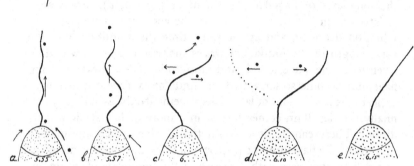

Fig. 25 A. Successive stages in the reduction in speed of the flagellar movement of a choanocyte of *Spongilla*. The collar is entirely retracted. The arrows indicate the direction of the water currents: the figures indicate the times of observation. *a*. Rapidly moving flagellum immediately after isolation of cell. *b*. Two minutes later, note increased amplitude of waves. *c*. Five minutes after isolation. *d*. Fifteen minutes after isolation. *e*. The flagellum at rest. Note the flexed position of rest. (After van Trigt.)

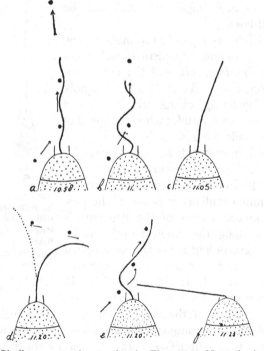

Fig. 25 B. Similar preparation to that in Fig. 25 A. Note the incompletely retracted collar. After 7 minutes the flagellum had come to rest (*c*), after shewing typical increase in the amplitude of its movements: 15 minutes later slow movement was resumed. Note pendular movement at *d*. Final cessation of movement with flexed flagellum is shewn at *f*. (After van Trigt.)

peculiar interest since they shew quite conclusively that the waves
pass round the flagellum of *Spirillum* as well as along it, thus
producing a torsional couple which rotates the animal about its
own axis in a direction opposite to that in which the waves

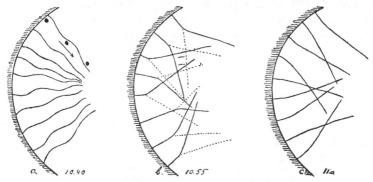

Fig. 26. Successive stages in the reduction in speed of movement of the flagella
of *Spongilla* inside an isolated flagellar chamber. Note the slow pendular movement in *b*, and the flexed condition of the motionless flagella in *c*. (After van
Trigt.)

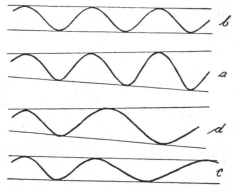

Fig. 27. Form of movement of *Spirillum volutans*. *a*. Note that the wave length
of successive loops is constant, but the 'amplitude' increases towards the
posterior end. *b*. Both wave length and amplitude is constant. *c*. The amplitude
is constant but the wave length increases towards the posterior end. *d*. Both
wave length and amplitude increase towards the posterior end. (After Fuhrmann.)

are travelling round the flagellum. In *Spirillum* the direction of
movement of the waves is always the same, but in *Vibrio* it can
be reversed so that the direction of movement and the direction
of rotation are also reversible.

In *Spirillum* the wave length may sometimes increase as the waves reach the extremity of the flagellum (see Fuhrmann, Fig. 27, *d* and *c*). The flagella of the flame cells of *Acanthocephalus* (Fig. 28) also appear to exhibit this phenomenon (Kaiser, 1893).

The flagella of sponges and of *Spirillum* obviously possess features in common and they may be regarded as typical examples of undulating movement in which a definite correlation exists

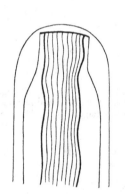

Fig. 28. Flame cell of *Acanthocephalus*. The wave length is said to increase as the waves pass along the flagellum. (After Kaiser.)

Fig. 29. *Bicosoeca lacustris*. *a*. The flagellum (*fl*) of an adult is uncoiling while the body emerges from the calyx. *b*. A young animal shewing the rigid and curved form of the flagellum. (After James Clark.)

between the amplitude and length of the waves on the one hand and the speed or frequency of movement on the other. Any attempt to analyse the nature of the motive mechanism must take cognisance of this fact (see p. 51).

The movements exhibited by some flagella are, however, more varied and less orderly than those of sponges or of *Spirillum*. Such variations are well illustrated within the group of the Flagellata (James Clark, 1868).

In *Bicosoeca gracilipes* the flagellum is not an undulating vibratile organ in the usual sense of the term, but extends upwards as a curved and rigid structure; only the distal portion is capable of the active jerky movements which direct particles towards the body of the animal. The proximal part of the flagellum only loses its rigid form when the body is withdrawn into the calyx. During such retraction the flagellum becomes coiled into a helix. As the body emerges from the bottom of the calyx, the flagellum slowly uncoils itself (see Fig. 29 a); then, after a few active vibrations, finally assumes its normal rigid appearance.

In *Anisonema concavum* (Fig. 30) there are two flagella. One of these (*fl*) is usually extended in front of the organism and is the main propulsive agent. It is a delicate actively vibratile structure and assists in the capture of food. The second flagellum is much longer and stouter than the first. It is of uniform thickness except at the base; it arises at the anterior end of the body but curves round and is directed posteriorly. It never shews any sign of undulatory vibration, but frequently moves from side to side, applying itself to solid objects and even acting as a lever by which the animal may lift its anterior end from the substratum.

The sensory functions of flagella are clearly seen in *Heteromastix proteiformis* (Fig. 31), which also has two flagella; one of these is always carried in front of the animal and its movements resemble those of the proboscis of an elephant, since it is constantly 'feeling' for obstructions and when one is found the flagellum seizes hold of it. The other flagellum trails behind the body and apparently acts both as a keel and as a rudder, since as the beast glides over the surface, the direction of movement can be changed by appropriate movements of this posterior flagellum. The motive power of *Heteromastix* is provided by the cilia.

Fig. 30. *Anisonema concavum*. Note the propulsive flagellum (*fl*) in front with waves more obvious at the tip than at the base. The larger flagellum (*fl*²) trails behind the animal and is not propulsive in function. (After James Clark.)

The general impression given by the Flagellate Protozoa is that the movements of the flagellum are often closely controlled by the animal. In sponges the flagellum either moves or is stationary. If it moves, the whole length of the flagellum takes part in an orderly series of events which are of the same type as long as the external environment remains constant. Among the Flagellata this is not the case, for the movement may vary in form and be restricted to certain regions of the flagellum in such a way as to produce an orderly purposive result although the movement itself is asymmetrical and irregular. Further in *Heteromastix* and *Peranema* the flagella appear to carry out sensory functions of a tactile nature at the same time retaining their contractile properties. In the metazoa these two functions are divorced from one another; most actively moving cilia are probably devoid of sensory functions although, in a few cases (*e.g.* the 'Stiftchenzellen' of Molluscs, and the 'bristles' found in invertebrate sense organs), cilia are no doubt sensory in function and are incapable of active movement. An interesting example of a sensory ciliated tract is seen in the grooves which lead from the aboral sense organ (see Fig. 83) of a Ctenophore to its locomotory cilia.

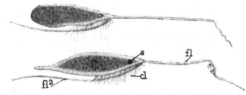

Fig. 31. *Heteromastix proteiformis.* Profile of the right side. One flagellum (*fl²*) the *gubernaculum* trails behind and underneath the animal, and acts as a keel and rudder. The other flagellum (*fl*) is extended in front and its tip exhibits irregular movements and is apparently of a sensory nature. The animal is propelled by means of cilia (*cl*). (After James Clark.)

It is possible to imagine that in the most primitive form the prolongations from the surface of a cell retain two fundamental properties (1) contraction, (2) the ability to detect mechanical vibrations and transmit their effects to other regions of the cell. Subsequent to this, the two functions are segregated from each other, thereby giving rise to cilia which are purely sensory or purely kinetic. In the latter case the degree of control which can be exercised over the contractile mechanism varies (see Chapter VII).

The cilium and flagellum as propellers.

From the description given of the movements executed by the frontal cilia of *Mytilus* it is not difficult to see how water is driven in the direction of the effective stroke and how, in the case of *Monas* swimming forwards, the animal is driven through the water in the opposite direction. The essential condition of these movements is that the work done on the water during the effective stroke is greater than that done in the opposite direction during the recovery stroke. During the effective stroke the whole length of the cilium is moving against the water, but during the recovery stroke the resistance is very greatly reduced owing to the tendency of the cilium to maintain its long axis parallel to the direction of movement. In the case of those cilia whose form during the recovery stroke is approximately the same as during the effective stroke it is doubtful whether any sustained directional work could be performed even if the effective stroke were more rapid than the recovery stroke.

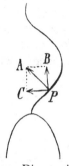

Fig. 32. Diagram illustrating Bütschli's theory of propulsion by a flagellum. A indicates the direction of the thrust exerted by the surface of the moving flagellum on the water. Bütschli believed propulsion to be effected by the component PB, whereas the rotatory movement was ascribed to the component PC.

Every propeller operates by driving astern some or all of the water which passes through it, and the reaction of this process furnishes a propelling force equal and opposite to the thrust of the propeller. If a ciliated organism is moving through still water with velocity V_1, the water which reaches the cilia can be regarded as travelling with the same velocity; after passing through the area operated on by the cilia this water will be travelling with an increased velocity V_2, so that immediately behind the organism is a current of water moving with velocity $(V_2 - V_1)$ in respect to the organism. This current is the propeller 'race.' If M is the mass of water passing through the ciliated region per second, the propelling force exerted by the cilia is $M(V_2 - V_1)$, and the direction in which it acts is opposite to the direction of movement of the race. Cilia can be regarded as paddles, and flagella as screws.

Of the numerous attempts to analyse the propulsive action of a flagellum, perhaps the best known is that of Bütschli (1887). His explanation was as follows. Any point (P) on the flagellum (see Fig. 32) carries out a series of lateral movements and in moving from side to side exerts a pressure on the water at right angles to its surface. If this force be PA, it can be resolved into two forces PB and PC. The force PB is directed parallel to the long axis of the organism and tends to drive it through the water cell foremost. The force PC is at right angles to PB and tends to rotate the organism on its own axis.

Hensen (1881) analysed the movement of a flagellate spermato-zoon as follows. " If the undulating membrane be examined, waves are seen to travel over its surface from the front to the back. Any one point moves laterally to and fro through a given distance with a force F. This force can be resolved into two components; one force kF is exerted tangentially to the membrane and tends to compress the latter; the other force k_1F is at right angles to the surface of the membrane. k_1F can in turn be resolved into two components, k_2F which exerts a backward pressure on the water and drives the organism forward, and k_3F at right angles to the surface which tends to rotate the organism on its own axis; opposed

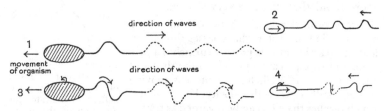

Fig. 33. Diagram illustrating the dependence of the direction of movement of an organism on the direction of the waves which pass along its flagellum. 1. The waves originate at the base of the flagellum and travel straight along it to its tip: the animal moves straight ahead with the flagellum pointing backwards. 2. The waves originate at the tip of the flagellum and pass to its base: the animal moves with the flagellum in front. 3. The waves originate at the base of the flagellum and pass along and round it in a clockwise direction: the animal moves with the flagellum behind and at the same time rotates in an anti-clock-wise direction. 4. The waves originate at the tip and pass backwards with an anti-clockwise rotation: the animal moves with the flagellum in front and rotates in a clockwise direction.

to this latter force is an equal and opposite force exerted by a point in which the transverse direction of movement is in the opposite phase."

Both Bütschli and Hensen appear to have overlooked the fact that in order to produce propulsion there must be a force which is always applied to the water in the same direction and which is independent of the phase of lateral movement. There can be little doubt that this condition is satisfied in flagellated organisms not because each particle of the flagellum is moving laterally to and fro but by the transmission of the waves from one end of the flagellum to the other, and because the direction of the transmission is always the same. A stationary wave, as apparently contemplated by Bütschli, could not effect propulsion since the forces acting on the water are equal and opposite during the two phases of movement. If however the waves are being transmitted in one direction only as is shewn in Fig. 33, definite propulsive forces are present which always act in a direction opposite to that of the waves. The propulsive power of the flagellum is equivalent to that which would be produced by projecting along the length of the flagellum a series of 'humps' of the same form as the waves, the velocity of the humps being made equal to the velocity of movement of each wave. If the waves pass from the base of the flagellum to its tip, the organism is driven forward in front of the flagellum; if the waves pass from the tip to the base the organism is drawn forward with the flagellum in front. If the waves pass along the flagellum in one plane there will be no force tending to rotate the animal on its axis: if, however, the waves pass round the flagellum as well as along it the organism will rotate. As the waves rotate, pressure is exerted on the water in such a way as to drive water in the direction in which the waves are travelling so that the flagellum with the attached cell tends to rotate in the opposite direction, just as when a cilium beats in one plane the movement of the organism is in the direction opposite to that of the effective stroke. If the waves pass clockwise round the flagellum, the organism will be rotated anticlockwise, as was observed by Riechert in

Fig. 34. Diagram illustrating the forces of rotation of *Spirillum*. The arrow z indicates the direction in which the waves travel round the flagellum. (After Riechert.)

Spirillum. "The body on account of the right-handed rotation of the flagellum (?), always turns to the left about its long axis."

In Fig. 34 each of the rotatory forces (e.g. x, y) generated by the waves on the flagellum can be resolved into a transverse component (e.g. ad), and the sum of these causes a rotation of the organism about its longitudinal axis vh.

The natural elasticity of cilia.

If the cilia on the frog's oesophagus or on the gills of *Mytilus* are brought to rest and one of the stationary cilia is then mechanically deformed by means of a needle, it very rapidly regains its normal shape when the needle is removed. Similarly if the tip of a moving cilium becomes entangled by an obstruction it is often deformed, but it regains its normal shape as soon as the tip is set free. In other words the cilium has considerable transverse elasticity. The question naturally arises—How far can the observed changes in the form of a beating cilium, such as those of *Mytilus*, be attributed to this elasticity? Carter (1924) has shewn that if a resting abfrontal cilium of *Mytilus* is moved through its backward

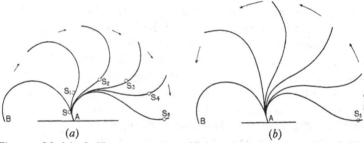

(a)   (b)

Fig. 35. Model of ciliary movement. *AB* is a strip of steel fixed at *A*. The strip is subjected to a bending force by the movable stop *S*; the bending force is applied at the base but gradually extends to the tip as the stop moves from *S* to S_5. On removing the stop the steel flies forward as in (*b*). Note that this model is of strictly limited value.

recovery stroke by means of a microdissecting needle the cilium is readily bent at the point of contact with the needle; as it passes forward by its own recoil, however, it is no longer readily deformed but is rigid, as is the case during the normal beat. It can be shewn that the characteristics of both recovery and effective strokes may be partially reproduced in a mechanical model. A curved strip of steel is firmly fixed in a horizontal position by one end *A*. Against the concave side of the steel is placed a movable stop *S*. By pressing the stop against the strip and at the same time moving it towards the free end of the strip, the latter moves back through the positions shewn in Fig. 35*a*. Under these conditions an increasing length of spring is kept in a state of tension; when the stop approaches the free end of the spring the latter takes on the form AS_5 in which it has the form of a sigmoid curve. If the stop is now removed the spring flies forward and the path traversed by its tip can be followed by placing, beneath the spring, a sheet of paper covered with a fine layer of soot. The form of the spring during the two phases of movement resemble those of cilia (Fig. 17) and suggests that the latter are storing potential energy during their recovery strokes by being stretched. A state of tension arises at the point of junction with the cell and gradually extends to the tip of the cilium; during the effective beat this energy is liberated and work is performed.

Like all mechanical conceptions of biological processes this analogy to a strip of steel must be used with caution. The rate at which energy is given out by a simple strip of steel depends solely on the extent to which it is stretched—the rate of stretching has little material effect. If, therefore, the steel were really an adequate model of a cilium we ought to find that the rate of the effective beat is independent of the rate of the recovery beat, and that if a resting cilium is moved through its complete recovery stroke by the mechanical pressure of a needle the cilium ought to fly forward, when released, with its normal powerful effective stroke. Neither of these conditions is satisfied by actual observation. The rate of the effective beat is always proportional to the rate of the recovery stroke (see Fig. 60), and the rate of elastic recoil from extraneous deformation is much slower than a normal effective stroke. The model of a steel spring is, however, of value and will be reconsidered at a later stage; for the moment it is convenient to mention certain alternative functions which have been associated with the natural elasticity of the cilium.

Heidenhain (1911) and later workers have regarded a cilium as consisting of a central elastic core with a contractile surface. It is suggested that the effective stroke is produced by the active contraction of the sheath on that side of the cilium facing the direction of the effective stroke, and that the cilium regains its resting position, when relaxation takes place, by virtue of the recoil of the elastic core. This view is undoubtedly attractive when we consider such cilia as those on the *latero-frontal* epithelium of *Mytilus* gills. In the resting position the cilium is perfectly straight and lies in the long axis of the cell: the effective beat appears to begin at the tip of the cilium and it is not easy to see how this can be effected by the cell itself. It is obvious that a system, whereby the elastic recoil of a cilium is used for carrying the cilium through its recovery stroke must be mechanically less efficient than the model described on p. 36. In the former case a quantity of energy proportional to the full curvature of the effective stroke is unavailable for useful work because it is stored in the cilium owing to its own elasticity. In the latter case, however, the energy available for work is only depleted in this way when the cilium passes its normal position of rest. Such pendular movements as already pointed out, are usually if not invariably restricted to cilia which do not perform much mechanical work but whose functions appear to be to deflect currents or particles from one direction to another.

Schäfer's hypothesis.

All attempts to determine the essential nature of ciliary movement are admittedly based on indirect evidence. One of the simplest schemes put forward is that of Schäfer (1891). His original hypothesis was formulated as follows: "If we suppose that a cilium is a hollow curved extension of the cell, occupied by hyaloplasm, and invested by a delicate elastic membrane, then it must follow that if there be a rhythmic flowing of hyaloplasm from the body of the cell, into and out of the cilium, an alternate extension and flexion of that process would thereby be brought about.... The same result might be got, supposing the cilium to be a straight and not a curved extension of the cell, if the enveloping membrane were thicker (or otherwise less extensible) along one side than along the other" (p. 198). Schäfer constructed a number of models to illustrate his hypothesis, and stated that "when the pressure is increased in the simple curved form, the artificial cilium straightens out, when diminished it bends over again; and by rhythmically repeating these movements a current is produced in any fluid in which the model is immersed, the direction of the current depending on the relative rate at which the increase or diminution of pressure within the cilium is brought about. If the increase of pressure be the more rapid, so that the artificial cilium straightens itself more quickly than, on relaxing the pressure, it bends over, a current is produced in the direction of the straightening; but if on the other hand, the increase of pressure be slow and the relaxation rapid, the current is then in the direction towards which the cilium bends" (p. 519, 1905). A spirited attack was made on this ingenious theory by Pütter (1902) who, like all other investigators, failed to detect any liquid axis in typical cilia. Schäfer maintained that "in all cilia or organs resembling cilia in which, on account of their size, it has been hitherto possible to observe any structure at all, they reveal themselves as tubular extensions of the cell substance enclosed by an elastic cuticula or pellicle." This statement is scientifically sound and dialectically admirable, but it does not carry us very far. No typical ciliary unit is, in point of fact, large enough to reveal any structure hollow or otherwise, and the only hollow extensions of the cell known to us are the tentacles of the Acinetarians which are not usually capable of doing work although they may be 'organs resembling cilia.' It is true that in *Asellicola digitata* Plate (1889) described the passage of water into and out of the tentacle several times a minute, but there is no evidence that any useful work was accomplished. On the other hand Bütschli (1887) maintained that the pump-like action of the tentacles was the means whereby the animal burst through the cuticle of its prey, although in *Ophryodendron* the tentacles are said to make rhythmical movements at their bases. It must probably be admitted that only very doubtful and indirect evidence supports the view that typical cilia are hollow.

The real value of Schäfer's hypothesis appears, however, to rest on two important facts: (i) the elasticity of the cilium plays an essential rôle, (ii) 'relaxation' as well as 'contraction' is an active process. Consider the simplest of Schäfer's models, viz. the hollow rubber tube which is curved when in the position of rest. If this tube is straightened mechanically and then released, it flies back to its position of rest at a speed independent of the rate at which it was straightened. It is behaving essentially the same as a bent strip of steel. If on the other hand water is slowly forced into the tube the latter is again straightened but this time the whole of the tube is stretched instead of only one side. The total energy available for work in this position is the energy stored in the walls of the model by virtue of their elasticity; the rate at which this energy is stored depends on the rate at which the water is pumped into the tube; the rate at which the energy is liberated depends on the rate at which the water is allowed to leave the tube.

Many authors, such as Pütter, Parker, Gurwitsch and Williams have rejected Schäfer's hypothesis and adhered to the suggestions of Heidenhain (1911) who, as already stated, attributed ciliary movement to the rhythmical contraction and relaxation of

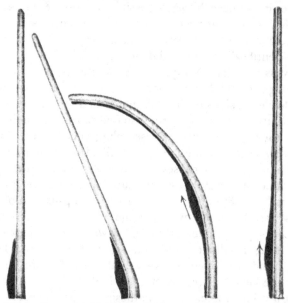

Fig. 36. Figure illustrating Heidenhain's theory of ciliary movement. One side of the cilium is regarded as contractile. The contractile area (in black) is restricted to the base in the case of pendular movement: in flexural movements the area of contraction moves along the cilium.

one side of the surface layer of the cilium itself. The central axis of the cilium is regarded by some authors as a mechanical elastic support for the active peripheral layer, whereas others regard the central axis as the 'contractile' unit, and the 'sheath' as the source of elastic recoil. Various modifications of this type of hypothesis have been advanced, but none of them have proved particularly fruitful.

The views adopted by Schäfer and by Heidenhain illustrate two radically opposite points of view. According to Schäfer the whole of the mechanical energy liberated by the cilium is derived as such from the cell and is transmitted as mechanical energy to be liberated as work by virtue of the elasticity of the cilium. According to Heidenhain, however, the mechanical energy liberated by the cilium is stored as chemical energy in the cilium itself until some disturbance arises in the cell or elsewhere which liberates it as mechanical energy. In other words, we may look on the cilium as a passive unit mechanically operated by the cell, or as an active structure potentially capable of autonomous contraction in all or some of its elements. From a morphological point of view these possibilities were considered by Williams (1907).

The cilium and flagellum as passive units moved by the cell.

The chief arguments in favour of the passive nature of vibratile organs appear to be as follows: (i) The flagella of *Spirillum* are only 0·05 μ in diameter (Fuhrmann 1910), and it is difficult to conceive how any heterogeneous mechanism of the type described in Chapter VI can be contained in a cross section which approaches ultramicroscopic dimensions. (ii) Many cilia and flagella shew no evidence of structural complexity even under the highest magnifications. (iii) Most cilia are motionless when separated from the cell. (iv) Some observers, including Williams (1907) claim to have seen active movements in the body of the cell corresponding in frequency to the movement of the cilia. These and other facts have led to the suggestion that the cilium or flagellum does not generate mechanical energy along its whole length but is merely a means whereby this type of energy, when generated at the distal end of the cell, is transmitted away from its source and is liberated for work. This view was in fact adopted by Henneguy (1898), Peter (1899), Benda (1899), Joseph (1903) and others.

To apply a series of rigid tests to this hypothesis is by no means easy, since the mechanical properties of such small vibratile structures have but imperfectly been considered. The following suggestions must therefore be received with caution; they are put forward in the hope that they may stimulate more adequate treatment.

The photograph of *Euglena* reproduced in Fig. 2 suggests that one type of flagellum has the properties of a thread which is very easily flexed, and which, at rest, shews little or no tendency to adopt a characteristic form. It is not difficult to see how such a structure might exhibit a series of undulatory movements if its base were moved laterally to one side and then to the other by contraction and relaxation of the protoplasm surrounding its base. If relaxation and contraction follow each other very rapidly such a flagellum would be thrown into a large number of loops of small 'amplitude': if the time which separates the end of relaxation and the beginning of contraction, and *vice versa*, is increased, the number of loops seen in the flagellum will be decreased, but they will be of larger 'amplitude.' It is possible to see how variations in undulatory movement, having a superficial resemblance to those described by van Trigt and others, can occur in a flexible inert flagellum if we assume, rashly perhaps[1], that sufficient kinetic energy can be stored in the flagellum to carry it for a reasonable time at a measurable velocity against the resistance of the water.

In order to use an undulating flexible unit as a means of propelling a body through the water in a line parallel to the axis of the undulations, it is essential that a force should be present which will cause the undulations to travel either away from the body or towards it; further, all waves must go in the same direction. The only way in which waves can be made to travel along a freely flexible unit is by establishing a state of tension. It is true that as long as a freely flexible thread is in motion a state of tension will exist owing to the 'drag' exerted by the surrounding water. Now any wave of distortion passing along such a filament will also meet with resistance from the water, and consequently the energy in the wave will decrease after leaving its source. If the figures given by van Trigt, Fuhrmann, and others are correct, little or no change in amplitude is seen in the waves which pass along a living flagellum, so we must assume that as a wave passes along a flagellum it gains energy as fast as it gives it up to the water.

Finally, we may consider flagella such as those of *Spongilla*, which undoubtedly have considerable natural rigidity. The movements which can be induced in an elastic rod by transverse distortions set up at one end are unfortunately difficult to analyse except in one special case. If the disturbances set up in one end of such a rod produce a distortion whose amplitude is small and whose form approximates to part of a sine wave, then it can be shewn that the distortion will propagate itself along the rod with a constant velocity (V) and without change in shape.

For a proof of this statement I am indebted to Mr L. B. Turner.

Let ABC (Fig. 37) be a distortion in which ABC is part of the sine curve

$$y = h \sin \frac{2\pi}{\lambda}.x.$$

If the amplitude (h) of the wave is small compared to its wave length (λ)

$$\frac{\delta y}{\delta x} = 0; \quad \frac{\delta^2}{\delta t^2}\left(\frac{\delta y}{\delta x}\right) = 0.$$

[1] See p. 90.

The element PP_1 of the rod at rest is bent to the position QQ_1.

$$PQ = y.$$

$$P_1Q_1 = y + \frac{\delta y}{\delta x}.\delta x.$$

The shearing force at Q is $\mathcal{S}$, and at Q_1 is $\mathcal{S} + \frac{\delta \mathcal{S}}{\delta x}.\delta x.$

Therefore the net force on the y-axis is $\frac{\delta \mathcal{S}}{\delta x}.\delta x.$

The acceleration of the element PP_1 on the y-axis is $\frac{\delta^2 y}{\delta t^2}$ and the mass is $m.\delta x$, where m is the mass per unit length of the rod.

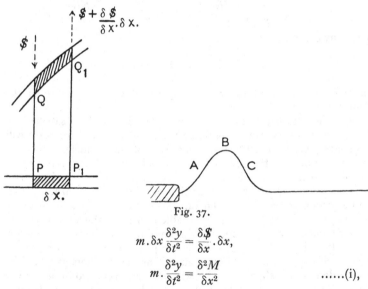

Fig. 37.

$$m.\delta x \frac{\delta^2 y}{\delta t^2} = \frac{\delta \mathcal{S}}{\delta x}.\delta x,$$

$$m.\frac{\delta^2 y}{\delta t^2} = \frac{\delta^2 M}{\delta x^2} \qquad \ldots\ldots(\text{i}),$$

where M is the bending moment at Q.

But $$M = -EK\frac{\delta^2 y}{\delta x^2},$$

where E is Young's modulus, and K is the moment of inertia about the neutral axis,

$$y = h \sin.\frac{2\pi}{\lambda}.x.$$

Therefore $$\frac{\delta^2 M}{\delta x^2} = EK\left(\frac{2\pi}{\lambda}\right)^2.\frac{\delta^2 y}{\delta x^2}.$$

Equation (i) therefore becomes

$$m \frac{\delta^2 y}{\delta t^2} = EK \left(\frac{2\pi}{\lambda} \right)^2 \cdot \frac{\delta^2 y}{\delta x^2},$$

$$\frac{\delta^2 y}{\delta t^2} = \left[\sqrt{\frac{EK}{m}} \cdot \frac{2\pi}{\lambda} \right]^2 \cdot \frac{\delta^2 y}{\delta x^2},$$

$$V = \frac{2\pi}{\lambda} \cdot \sqrt{\frac{EK}{m}}.$$

If a number of such distortions pass down the filament with frequency n per second, if the wave length of each is λ and there is no pause between successive waves then $n\lambda = V$, so that

$$\lambda = \sqrt{\frac{2\pi}{n} \cdot \sqrt{\frac{EK}{m}}} \quad \text{and} \quad V = \sqrt{2\pi n \cdot \sqrt{\frac{KE}{m}}},$$

or for any given filament

$$\lambda = C \sqrt{\frac{1}{n}} \quad \text{and} \quad V = C \sqrt{n},$$

where C is a constant determined by the physical structure of the filament.

Regarding the flagellum as a passive elastic rod which satisfies the above conditions we can see how the wave length of a disturbance will be inversely proportional to the square root of the frequency of vibrations set up in it by the cell, and that the velocity with which the waves will travel along the flagellum will be directly proportional to the square root of the frequency. These conclusions appear to be in harmony with the behaviour of the waves seen on a flagellum, but it is unfortunate that the physical analysis is only applicable to flagella which are operating at a high enough speed to reduce the amplitude of the waves to a low value. If it be assumed, that for all amplitudes the wave length and velocity of propagation are functions of the same factors as when the amplitude is small, then it can be seen that an elastic rod operated at one end can act as a propulsive element as long as no reflected waves exist.

There appear to be two ways in which to test the conception of the flagellum as an inert but flexible rod. If a wave can pass down a flagellum at constant speed and without change in form, then the wave must be part of a sine wave (otherwise as it moves along it will resolve itself into a series of sine waves of different wave lengths each travelling with a different velocity, so that the form of the wave will vary at different points along its path). Now in order to produce a sine wave in an elastic rod it is necessary to exert a bending force and a longitudinal thrust, and it is very difficult to see how such forces could be exerted by movements entirely restricted to that part of the flagellum which is in contact with the cell. Secondly, if a vibrating rod is to exhibit no reflected

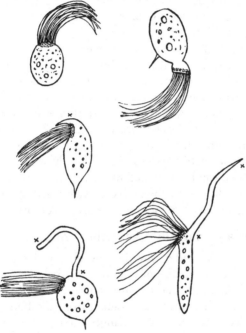

Fig. 38. Formation of contractile lobopodia from distal end of isolated ciliated cells of *Unio*. The lobopodia are marked ×, and on being detached from the cell are capable of active movement whereas the cilia themselves cease to move. (After Merton.)

waves which travel back towards their original source, the vibrations must be very highly damped; but if the damping is high, the amplitude of the waves must invariably decrease as they pass from their source, and this is apparently not the case in living flagella.

The evidence, as far as it goes, seems to suggest that a flagellum cannot be regarded as an inert flexible unit operated mechanically by the cell since waves of varying form pass along its length without change in shape and without apparent change in amplitude.

That movements of a flagellum or cilium are not infrequently accompanied by mechanical movements within the cell is highly probable, but this evidence does not shew that the flagellum or cilium is not also an actively contractile element. The contractile properties at the distal end of ciliated cells have recently been investigated by Merton (1924, 1927).

If the ciliated gills of *Unio* or *Anadonta* are resolved into their constituent cells, the latter move about actively by means of their cilia. In many cases isolated cells push out contractile processes or 'lobopodia' which, on becoming detached from the cell, exhibit prolonged autonomous movement. Merton states that when all this distal 'kinoplasm' has left the cell, the cilia cease to beat[1]. One is inclined to think that both cilia and distal protoplasm are part of an active contractile system. In a flagellum the whole length of the flagellum can be imagined to retain its contractile functions, whereas in pendular cilia the active elements are restricted to the base, leaving the distal region as a passive structure.

The cilium and flagellum as active units.

We may now consider how far the observed facts harmonise with the view that the cilium is to be regarded as an active living unit. By this is meant that the mechanical energy liberated by the vibratile organ is not derived, as such, from the body of the cell, but is derived from chemical changes taking place in the organ itself. In the first place, the small transverse dimensions of some flagella are not completely irreconcilable with such a view; when we remember the biological complexity of a single chromosome, the potentiality for structure in a flagellum seems considerable. Although a chromosome does not, as far as we know, expend considerable amounts of energy which must continuously be replaced ten or twelve times a second along its whole length, yet there is no doubt that all the main phenomena of life can be exhibited by organisms of extremely small dimensions. Further, optical homogeneity of protoplasm is the rule rather than the exception in

[1] Merton's observations have recently been criticised by v. Rényi (1926); see also Merton (1927).

nature. The mere absence, therefore, of a visible structure in a flagellum or cilium does not necessarily prove that it is not an active living unit.

One of the most direct arguments in favour of the passive nature of the cilium rests on the statements of Kraft (1890), Williams (1907), and others, to the effect that protoplasmic movements can actually be seen in the distal end of the cell. On the other hand, such observations are distinctly uncommon, for in most cases no such movements can be seen. It is conceivable that where protoplasmic movement is visible it is the mechanical effect produced by an active cilium and not the cause of movement in the latter. On the other hand, as already suggested, both the vibratile element and the protoplasm may be active parts of one living contractile machine.

Whereas there is no doubt that cilia are incapable of prolonged movement when they are detached from the body of the cell, there are not a few cases recorded in which movement can be seen for a short period after excision. Klebs (1881) reported movement in the isolated flagellum of *Trachelomonas*; Bütschli (1885) described in some detail movements of a detached flagellum of *Glenodinium cinctum*, this organ carrying out distinct 'corkscrew' movements for about a minute after excision. Other observations of this nature were made by Schilling (1891), Fischer (1894) and Rothert (1894). Among the metazoa evidence of autonomous movement may be derived from the ciliated plates of the Ctenophore *Pleurobrachia*, since on teasing off one of the cilia from a plate distinct twitchings can be observed in the fibrils at the proximal end of the cilium.

Almost overwhelming evidence in favour of the active nature of the cilium or flagellum is provided by the following facts.

(i) In some cilia (see Fig. 12 b) there can be no doubt that the angular velocity of the tip is greater than that of the base, so that the whole cilium becomes bent during the effective stroke. It is exceedingly difficult to see how such a movement could be effected by the liberation of kinetic energy at the extreme base of the cilium[1]; on the other hand, the nature of this movement can be readily visualised

[1] If one side of a cilium is assumed to be less extensible at the base than at the tip, Schäfer's hypothesis can be harmonised with the facts.

by the assumption that the effective stroke is due to the development of a bending force along the whole length of the cilium.
A straight elastic rod when subjected to a uniform bending force
along its whole length acquires the form of an arc of a circle,
the radius of which is inversely proportional to the bending force.
Fig. 39 indicates the types of movement which can be effected in
a straight rod which is subjected to bending forces applied either
simultaneously over its whole length or which spread along the
rod from tip to base.

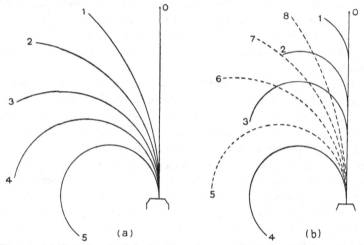

Fig. 39. (a) Straight elastic rod subjected to a uniform bending force simultaneously along its whole length. If the bending force is removed simultaneously
along the whole rod, the form of recovery is the same as that during bending.
(b) Bending force developing at the tip and being transmitted to the base, 1–4;
if the bending force is removed simultaneously along the whole length, the
recovery stroke, 5–8, differs from the bending stroke. Note that during relaxation the rod forms an arc of a circle of increasing radius of curvature.

We have already seen that the movements of the frontal cilia
of *Mytilus* are explicable on the assumption that the bending force
starts at the base and is subsequently transmitted to the tip.

(ii) Among the Mastigophora it is not at all uncommon for the
flagellum to extend forward in front of the cell and to move the
animal through the water by movements restricted to its tip. This
type of movement is seen, for example, in *Peranema* and in *Monas*.
It hardly seems possible to believe that sufficient kinetic energy

can be transmitted along the proximal part of the cilium to induce active movements at the tip. Dr Bidder informs me that he has seen, not infrequently, lateral movement of the extreme tip of a sponge flagellum although the rest of the flagellum remained motionless and straight. These facts together with the observation that sudden bends may occur at any point on a slowly moving flagellum can only be explained by the assumption that kinetic energy can be generated at any point in the flagellum.

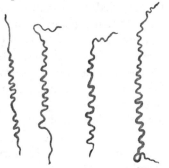

Fig. 40. *Treponema pallidum*. Note the irregularity in the waves which appear to die out towards the two ends of the animal.

Fig. 41. *Spirochaeta plicatalis*. Note the very large number of waves of varying length and amplitude. (After Doflein.)

(iii) As already mentioned there is no evidence that all waves of distortion passing along a flagellum undergo changes in amplitude. If this could be supported by quantitative measurement quite conclusive evidence would be available to show that the flagellum is an active unit capable of generating kinetic energy along its length. A moving wave cannot provide the energy for propelling an organism and at the same time pass on with unreduced amplitude unless the energy being lost is continually being replaced as the wave moves along.

(iv) In all tractella, or flagella which draw an organism forward through the water, the waves of distortion pass from the tip of the flagellum to its base. It hardly seems possible to imagine that such movements could be exhibited by an inert filament whose sole source of energy lay in the body of the cell.

We may therefore conclude that a flagellum or a cilium is to be regarded as an active unit (comparable perhaps to a muscle fibre) in that it generates its own mechanical energy. The scheme put forward by Heidenhain (1911), suggests that along the cilium there is an area of contractile protoplasm in which rhythmical changes of longitudinal tension effect a transverse bending of the whole cilium. It is, however, difficult to see how such a system could allow a single cilium or flagellum to carry out movements of a variable nature: if the path of contraction is fixed the form of movement is strictly limited.

Fig. 42. Photograph of a preparation of *Spirillum volutans*. The successive waves do not appear to differ materially in amplitude or wave length. (From Fuhrmann.)

Riechert pointed out that variations of wave length would result if we postulate a variation in the relative rate at which a contraction can pass along a flagellum in a longitudinal and in a transverse direction. It is not clear that the contraction of the ciliary sheath (postulated by Heidenhain) is necessarily restricted to the development of forces in a longitudinal direction. When a stimulated muscle fibre develops a tension, the force of the contraction is directed along the longitudinal axis of the fibre; we do not know how this force is produced, but if it can be produced longitudinally in a muscle fibre, it is not unreasonable to suppose that it can be produced transversely and also as torsion in a flagellum or cilium. Consider, for example, an india-rubber rod. If its ends are compressed it shortens and a longitudinal strain exists between adjacent particles. If it is bent, a transverse strain is set up. If one end is fixed and the other is twisted, torsional strains arise. We can imagine that in a muscle a longitudinal strain is set up internally and shortening is the result. If a corresponding transverse strain is set up internally in a cilium it will cause bending; and a torsional strain will cause twisting. In other words the essential changes in a muscle fibre and in a flagellum may very likely be the same, although they occur along different axes.

As Heidenhain himself pointed out, if a cilium bends because a longitudinal tension is set up along one side, the nearer the contractile elements are to the surface of the cilium the more powerful is their effect, and for this reason he postulated a contractile sheath and an inert elastic filament in the centre. Duboscq (1907), however, gave fairly strong evidence to shew that this is not a true picture of the flagellated spermatozoa of *Paludina*. When these large spermatozoa are placed in water they swell up and although the central filament of the tail becomes totally enclosed by protoplasm, it still continues to carry out active movements. It seems more reasonable to suppose that whatever be the machinery of a muscle fibre, it exists also in a cilium or flagellum but is orientated in a different way.

If a cilium or flagellum is to be looked upon as equivalent to a muscle fibre then its failure to move when isolated from the cell may be due to lack of an appropriate stimulus; or in those cases where transitory movement does occur in the isolated state we may imagine that the machine fails for lack of a renewal of its potential energy. The phenomena of muscular excitability, transmission, contraction and tone provide ample scope for ciliary speculation, but until means are found of subjecting the problem to experimental analysis any discussion can have very little value.

Variation in the form of active flagella which exhibit regular undulatory movements only.

In order to see how variations can occur in the form of the waves passing down an active flagellum it is necessary to consider two distinct types of movement: (*a*) where the waves pass along the flagellum causing it to execute movements in one plane only; (*b*) where the waves pass along and round the flagellum, causing each element to travel in a circular or elliptical orbit. In the first case we may imagine that as soon as a disturbance reaches a length of the flagellum which is at rest there is generated in this length a force which causes the length to bend into an arc of definite radius. If, in Fig. 43, *ADB* is a length of flagellum which is at rest, and a uniform bending force (*f*) is generated along its length, then *AB* will bend into an arc of a circle *AGB* and be in equilibrium when the radius of curvature (ρ) satisfies the equation

$$f = \frac{EK}{\rho},$$

where *E* is Young's modulus, *K* is the moment of inertia about the neutral axis, and ρ is the radius of curvature. If *AGB* is the position of equilibrium, then the radius of curvature OB will be inversely proportional to the bending force, and if the maximum displacement (DG) is not very large it can be expressed in terms of the length AB. If CG is the diameter of curvature through G,

$$CD \times DG = AD^2.$$

Let $OC = \rho$; $DG = h$; $AB = L$,

$$h\,(2\rho - h) = \left(\frac{L}{2}\right)^2.$$

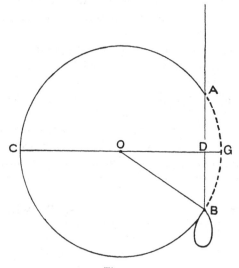

Fig. 43.

Since h is small compared to 2ρ,

$$h = \frac{L^2}{8\rho} = \frac{fL^2}{8EK}.$$

If the bending force is constant, the 'amplitude' of the bend varies directly as the square of the length being bent. For larger amplitudes this approximation will not apply, but the amplitude will still increase more rapidly than the length. This is shown graphically in Fig. 44, where the amplitude of the longer arc ABC is arbitrarily taken as equal to the radius of curvature; on reducing the length of the arc to one-half of ABC (*i.e.* to ADE), the amplitude DG is reduced to one-third of BF, although the radius of curvature remains the same. It is therefore possible to account for the large amplitude characteristic of flagella which are moving slowly in one plane if we assume that the bending force is less rapidly removed by relaxation than is the case during faster movements. At the same time we must assume that the velocity at which the bending wave passes along the flagellum remains more or less constant.

A bent flagellum will, however, only lie in one plane if the plane of bending coincides with the longitudinal axis of the flagellum and if torsional forces are absent. If the plane of bending is inclined at an angle to the longitudinal axis or if the flagellum becomes twisted when a bending force is generated uniformly along its length, the flagellum will acquire the form of a regular helix, whose form will depend on the physical nature of the flagellum and on the forces generated. If the

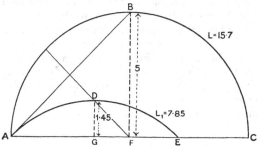

Fig. 44. Diagram illustrating a possible way in which the 'amplitude' of a flagellar wave may depend on the length of the flagellum subjected to a uniform bending movement.

bending force per unit section be f, then the radius of flexural curvature (ρ) will be $\dfrac{EK}{f}$. If the plane of bending is inclined at an angle (ϕ) to the longitudinal axis of the flagellum, then the diameter of the generating cylinder of the helix is 2ρ and the pitch of the helix is $2\pi\rho \tan(90° - \phi)$. In terms of flagellar movement these two quantities express the amplitude and the wave length of the distortions seen during movement. Since ρ is inversely proportional to f and ϕ is constant, it follows that the greater the bending force generated the smaller is the apparent amplitude and wave length seen in the distorted flagellum. On the other hand, if the plane of bending coincides with the longitudinal axis of the flagellum and the helical form is the result of torsional forces, then for a given bending force $\left(\dfrac{EK}{\rho}\right)$, the diameter of the generating cylinder of the helix is $2\rho \cos^2 a$, and the pitch of the helix is $2\pi\rho \sin a \cos a$, where a is the pitch angle of the helix. As long as a is constant the amplitude and wave length are again inversely proportional to the bending forces generated in the flagellum.

In the above discussion two assumptions have been made. It is assumed that the mechanical activity (or some derivative therefrom) in one element of the flagellum provides the stimulus which activates the element next to it. This seems a reasonable view, since each wave seems to travel regularly along the flagellum. Secondly, it is assumed that when a wave passes along and round the flagellum it does so slowly in comparison with the speed at which the force of flexure rises to its full intensity when once an element is stimulated. It is, of course, more likely that the bending wave will travel a finite distance before these forces reach a maximum. The result of this would be, in the case of simple flexural movement, to make the distortions approximate in form to that of a sine wave[1] rather than to the arc of a circle. If the bending is accompanied by torsion and the latter is uniformly distributed at all points in the wave, the cross-section of the generating figure of the spiral will no longer be circular but will approximate to an ellipse. If the torsional force varies with the flexural force the cross-section of the figure may be circular. The point is of considerable importance since, according to van Trigt, the tip of a flagellum in *Spongilla* travels along an ellipse. Again, the tip of a slowly moving flagellum does not appear to travel in a circular orbit of wide diameter as would be the case in the uniform systems described above.

Hypothesis of ciliary movement.

Schäfer's hypothesis of ciliary movement (see p. 38) is based on the suggestion that certain units in the vibratile structure expand under hydrostatic pressure more than others and thereby cause a distortion of the whole cilium. When the pressure is released the energy stored in the stretched unit restores the cilium to its original position. The weakness of the hypothesis lies in the suggestion that the vibratile organs are hollow structures whose cavities extend from the base to the tip.

From the evidence already given we may conclude that the mechanical energy for movement is generated along the length of the cilium and is not transmitted as hydrostatic pressure by the cell. We know that when water is removed from the cilium by

[1] It may be noticed that whereas a sine-wave is the only type of distortion which can be propagated unchanged along a passive flexible element, a wave of any form can be propagated unchanged along a uniform 'living' filament since the energy of the wave is the same in quantity and distribution for all positions of the wave.

plasmolysis, the cilia of *Mytilus* or of the frog come to rest at the beginning of their preparatory stroke (Gray, 1922), so that there is some slight experimental evidence in favour of regarding the

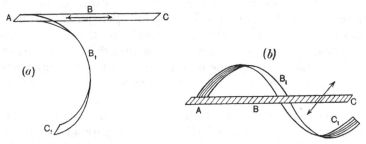

Fig. 45. Figures shewing the form of a strip of paper caused to bend by adding water to one side. In (*a*) the strip *ABC* has been cut so that the natural axis of bending (indicated by the arrow) is parallel to the longitudinal axis of the strip and the latter bends into an arc of a circle AB_1C_1. In (*b*) the strip has been cut so that the axis of bending is inclined at an angle to the longitudinal axis of the strip; the latter bends into a helix. Suitable strips can be cut from a sheet of note-paper; type (*a*) is provided by a strip cut parallel to the shorter edge of the paper; type (*b*) by a strip cut along a diagonal of the sheet.

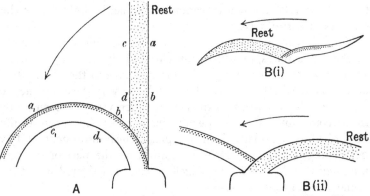

Fig. 46. Hypothetical mechanism of different types of cilia. *A*. The cilium at rest is straight, and the water molecules are uniformly distributed. Contraction occurs by the aggregation of water molecules to one side of the cilium, causing this side to become convex. Relaxation occurs when the water molecules again distribute themselves uniformly. *B* (i). The cilium is at rest at the end of the effective stroke and in this position the water molecules are uniformly distributed. The preparatory stroke is complete when the water is aggregated to one side. *B* (ii). The cilium is at rest at the beginning of the effective stroke and during the effective stroke the water is aggregated to one side.

normal stroke as due to a redistribution of water within the cilium. If one side of a straight cilium becomes capable of absorbing more water than the other, the cilium will bend into the arc of a circle whose convex side is the side containing the excess of water. A model can readily be made with a strip of paper (Fig. 45). If both sides of the strip are equally dry or equally wet the strip remains flat, but if one side becomes more moist than the other the paper bends so that the damp side is outermost. As the moisture soaks through the paper the strip gradually straightens. We may picture a cilium as in Fig. 46. Whilst at rest, water is equally distributed across its whole section, movement occurs because certain elements lying nearer to the side AB than to CD acquire an affinity for water, thereby causing water to flow from the side CD to AB; AB expands and CD contracts—thereby bending the cilium. Relaxation occurs because AB loses its increased affinity and the water distributes itself equally over the cross-section of the cilium. The essential change in the affinity for water may well be due to a modification in the degree of ionisation of a protein or similar molecule: thus, the localised production of an acid would produce this effect whilst the neutralisation of the acid would return the system to the *status quo*.

The following points in this scheme are noticeable: (i) the general mechanism is that applicable to muscle (see Hill and Hartree, 1920), since an essential feature is a localised change in the distribution of water; (ii) both effective beat and recovery beat are active processes which depend on the rate at which water is taken up by the activated units and the rate at which it is lost; (iii) it can readily be applied to pendular cilia or to typical helical movement, and by postulating a system of localised control on the part of the cell, a limited length or area of a flagellum can be activated whilst the remainder continues at rest.

Position of ciliary rest.

In the case of locomotory cilia under the control of the organism it is customary to find that the cilia come to rest in the position from which the effective stroke begins. This is well seen in Cteno- phores. In Fig. 47, position A indicates the position of rest. As soon as the cilium begins to move it passes through the form shewn at *B* and *C*, while the recovery stroke returns the cilium to position *A*. Other examples of this type are provided by the velar cilia of many veliger larvae, by the velar cilia of *Vorticella* and by the latero-frontal cilia of *Mytilus*. On the other hand, in the case of the velar cilia described by Williams (1907), it is clear that the position of rest lies in front of the position at which the effective stroke begins.

In the case of cilia which normally beat without interruption during the life of the animal (*e.g.* the frontal cilia of *Mytilus*, or those on the oesophagus of the frog) it is difficult to define any

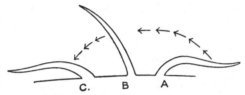

Fig. 47. Movement of ciliated plate of a Ctenophore. *A* is the position of rest at the end of the backward or recovery stroke. *C* is the position at the end of the effective stroke. The arrows indicate the direction of the effective stroke.

natural position of rest. Such cilia can, without any apparent permanent injury, be brought to rest by a variety of methods at the end of the effective stroke—that is, in exactly the opposite position to that seen in Ctenophores.

Reversal of ciliary movement.

The reversal of cilia in metazoa is an exceedingly rare phenomenon, and in no case has a detailed account been given of the movements executed by a single cilium. The literature is briefly reviewed by Parker (1905), who quotes the following instances of reversal: on the labial palps of *Mytilus* (Purkinje and Valentin, 1835, Engelmann, 1868); sponges (Miklucho-Maclay, 1868); planarians (Minot, 1877, Iijuma, 1884); actinians (Parker, 1896, Vignon, 1901, Torrey, 1904).

By far the most convincing description is that given by Parker (1896) of the cilia on the lips of the sea-anemone, *Metridium marginatum* (Fig. 48). If small particles of carmine are placed on the labial zone of an expanded anemone they move outwards towards the tentacles. If however a fragment of crab's muscle or other nutritive substance is used instead of carmine it is carried into the mouth and not outwards. That these phenomena are due to a reversal of the ciliary stroke was confirmed by the following experiment. "When a piece of the lip is placed in sea-water under a microscope, the normal direction of the ciliary stroke can be easily observed. If, now, such a preparation is flooded with sea-water containing meat juice an immediate reversal can be seen to take place, and on washing the preparation subsequently with pure sea-water, a return to the normal direction occurs Such changes can be repeated many times without impairing the piece of tissue. Hence there can be no doubt that the dissolved substances in the meat

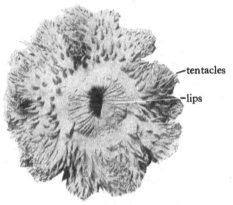

Fig. 48. Vertical view of the oral disc of *Metridium*. (After Parker.)

cause ciliary reversal" (Parker, 1905). The extent of the area over which the reversal of cilia takes place was demonstrated as follows. A quantity of carmine was discharged over the lips of a *Metridium* and the cilia at once began to move them away from the mouth. If, during this operation, a piece of meat was dropped on the lips, it was seen that although the cilia immediately round the meat reversed their movement, those in front, behind, and at either side of it continue to wave outwards, the area of reversal advancing in front and dying out behind the particle as the latter approaches the mouth. Reversal is therefore a localised phenomenon and only lasts as long as the stimulating substance is present.

Parker (1905) has shewn that reversal of the labial cilia of *Metridium marginatum* also occurs if the tissue is exposed to a

2½ per cent. solution of potassium chloride in sea-water, and that both the metachronal wave and the effective strokes of the cilia are reversed. Reversal never occurs in the cilia on the tentacles nor on the siphonoglyph. Working with another species of anemone, *M. davisi*, Torrey (1904) observed ciliary reversal in response to such

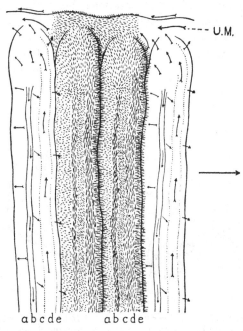

Fig. 49. Enlarged semi-diagrammatic figure of the palp folds of *Ostrea*, shewing the direction of ciliary currents. *a–e* indicate the five tracts of cilia on exposed surface of the folds. *U.M.* Upper margin: the large arrow on the right of the figure indicates the direction of the mouth. Note the close proximity of the different currents: there is no evidence to support the view that any individual current is ever reversed in direction. (After Yonge.)

inert substances as paraffin, glass, and paper, and concluded that the phenomenon is due to some type of mechanical disturbance.

The evidence afforded by these facts is strongly in favour of a reversal in the direction of the effective beat, but it is not absolutely conclusive. There can be little doubt that the older statements concerning the labial palps of Molluscs are wrong, since all recent workers agree that on these organs there are two permanent ciliated tracts which beat in opposite directions and which do not reverse their movements (Kellogg, 1915, Yonge 1926). Yonge, describes no less than five separate ciliated tracts on the exposed distal surfaces of the palp folds in *Ostrea* (see Fig. 49). The strongest exhalent system of cilia lies, however, in the furrows between the folds. Wallengren, who first described these ciliated tracts, was of opinion that muscular movements determined whether particles came into contact with the orally or aborally directed cilia. This view was supported by Allen (1914, 1921) and by Nelson (1924). Nelson states that the rejection of food particles from the region of

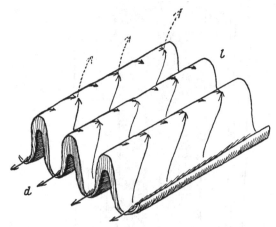

Fig. 50. Ciliary tracts in the liver diverticulae of *Helix*. Note that the directions of the currents are different in the furrows to those on the ridges. (After Merton.)

the mouth is due to reflex erection of the ridges of the palps which brings into play groups of cilia which beat away from the mouth. By narcotising the muscles with magnesium sulphate the filaments of the palps failed to erect, and a continuous supply of particles was observed passing to the mouth.

Such parallel but opposing ciliary tracts are quite common in Molluscs, Ctenophores and other animals, and are often disposed at the crests and furrows of a ridged surface (Fig. 50). Since the oral disc of *Metridium* is ridged and its muscles are extremely sensitive to mechanical stimulation, one would like to be quite certain that the reversal of the currents observed by

Parker is not due to two separate series of cilia which beat in opposite directions on the ridges and in the furrows. That there are material grounds for this criticism is shewn by the following observations of Elmhirst (1925) on *Actinoloba dianthus*. "Longitudinal grooves run down the gullet, and when food is being swallowed the inflow is along the grooves; conversely

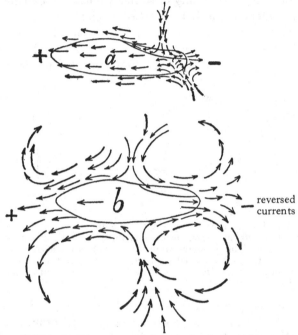

Fig. 51. Reversal of the water currents of *Paramecium* induced by an electric field. The currents are reversed at the anterior end which is facing the cathode. (After Jennings.)

a ciliary outflow runs up the ridges, for example, when a bolus of waste is discharged it is passed out by the cilia on the ridges aided by a certain amount of contraction of the stomodaeal wall. At times there is a vortex in the gullet when both sets of cilia are in action at once."

Only in the protozoa has a reversed beat been seen in individual cilia, and even in this group the data are few. In investigating the effect of induction shocks on *Paramecium*, Statkewitsch (1904) observed reversal of the cilia nearest to the anode when the movements of the animal were slowed down in a viscous solution (Fig. 51). The detailed form of the cilium is not described, but Jennings (1915) states that they 'strike forward instead of backward.' We have seen that during backward movement the form of the flagellum in *Monas* is totally different from that during forward progression; at the same time the ability of so many ciliates to execute backward movements can hardly be ascribed to such changes. That reversed currents occur when a ciliate swims backwards is of course obvious, but at present we can only speculate as to the mechanism whereby these are produced. A careful study of the behaviour of individual

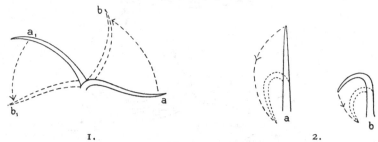

Fig. 52. Hypothetical scheme whereby the direction of a ciliary current could be reversed without altering the essential direction of the effective stroke. 1. The 'tone' of the cilium is altered from a to a_1 so that the effective stroke changes from $a \to b$ to $a_1 \to b_1$. The effect of KCl on the latero-frontal cilia of *Mytilus* is shewn in 2. Note change in 'tone'; a is the beat in normal sea-water, b is the beat in the presence of KCl.

cilia in the protozoa is much to be desired, as it would throw considerable light on the whole ciliary mechanism.

It has already been mentioned that Parker (1905) induced a reversal of the cilia of *Metridium* by the addition of $2\frac{1}{2}$ per cent. potassium chloride to sea-water, and recently Mast and Nadler (1926) have observed that this salt induces reversed movement in *Paramecium*. As a rule, this reagent simply increases the speed of ciliary beats (Gray, 1921), but in certain cases the form of the beat is altered; the *latero-frontal* cilia of *Mytilus* are thrown by potassium chloride into a state of prolonged contraction, recalling the tonic contraction of a muscle (see p. 99). The cilia are not stopped by the reagent, but vibrate rapidly with greatly reduced amplitude. In cilia disposed along a ridge of tissue and beating along the ridge, tonic contraction might reasonably lead to a reversal of the current although there would be no inversion of effective and recovery strokes (Fig. 52). At the same time, it is probable that KCl would induce contraction of the muscles if these organs are present.

It is difficult to imagine how the frontal cilia of *Mytilus* or flagella such as those of *Monas*, could perform any appreciable amount of work during their recovery strokes; but if a cilium is of such a type that there is not much difference between the form of the two strokes it is conceivable that the nett effect of the beat could be reversed by quickening the recovery stroke and slowing the effective stroke as appears to be the case in some protozoa. Parker (1906) suggested that the reversal of *Metridium* cilia was effected by a system of flexor and extensor elements, and his view was elaborated by Williams (1907).

References

(Those of general interest are marked with an asterisk.)

Allen, W. R. (1914). Biol. Bull. 27, *127*.

—— (1921). Biol. Bull. 40, *210*.

Becker, O. (1857). Molescholt's Untersuch. 2, *71*—[from Pütter].

Benda, C. (1901). Arch. f. Anat. u. Physiol. Suppl. *147*.

*Bidder, G. P. (1923). Q.J.M.S. 67, *293*.

Bütschli, O. (1887). Bronn's *Klassen und Ordnungen*. *Protozoa* 3. Leipzig.

Carter, G. S. (1924). Proc. Roy. Soc. 96 B, *115*.

Dellinger, O. P. (1909). Journ. Morph. 20, *171*.

Duboscq, O. (1907). Comptes rend. Assoc. d'Anatom. 9.

Elmhirst, R. (1925). Scottish Naturalist, *149*.

*Fuhrmann, F. (1910). Centr. f. Bakter. II, 25, *129*.

Gray, J. (1922). Proc. Roy. Soc. 93 B, *104*.

*Gurwitsch, A. (1904). *Morphologie und Biologie der Zelle*. Jena, 1904.

*Heidenhain, M. (1911). *Plasma und Zelle*, 1, 2. Jena.

Hensen, V. (1881). Hermann's *Handbuch der Physiolog.*, Bd. VI, *89*.

Hill, A. V. and Hartree, W. (1920). Phil. Trans. Roy. Soc. 210 B, *153*.

Iijuma, I. (1884). Zeit. f. wiss. Zool. 40, *359*.

*James-Clark, H. (1868). Annals of Nat. Hist. 4th series, vol. I.

Jennings, H. S. (1915). *Behaviour of the lower animals*. New York.

Kaiser, J. E. (1893). Biblioth. Zoologica, Bd. 2.

Kellogg, J. L. (1915). Journ. of Morph. 26, *625*.

*Kraft, H. (1890). Pflüger's Archiv, 47, *196*.

*Krijgsman, B. J. (1925). Archiv f. Protisk. 52, *478*.

Mast, S. O. and Nadler, J. E. (1926). Journ. of Morph. 43, *105*.

Merton, H. (1924). Zeit. f. Wiss. Biol. (Zellen und Gewebelehre), 1, *671*.

—— (1927). Zeit. f. Anat. und Entwickl. 83, *222*.

Minchin, E. A. (1912). *Introduction to the study of the Protozoa.* London.

Minot, C. S. (1877). Arb. zool. zootom. Inst. Würzburg, 3, *405*.

Miklucho-Maclay, N. (1868). Zeit. f. Naturwiss. Jena, 4, *221*.

Nelson, T. C. (1924). Proc. Soc. Exp. Biol. and Med. 21, *166*.

Parker, G. P. (1905). Amer. Journ. Physiol. 13, *1*.

Perrin, W. S. (1906). Arch. Protisk. 7, *131*.

Plate, L. (1889). Zool. Jahrb. 3, *135*.

*Pütter, A. (1902). Erg. d. Physiol. 2, Abt. 2, *1*.

v. Rényi, G. (1926). Zeit. f. d. gesamte Anat., Abt. 1, 81, *691*.

*Riechert, P. (1909). Zentr. f. Bakt., Abt. 1, 51, *14*.

*Saguchi, S. (1917). Journ. Morph. 29, *217*.

Schäfer, E. A. (1891). Proc. Roy. Soc. 49, *193*.

*—— (1904). Anat. Anz. 24, *497*.

*—— (1905). Anat. Anz. 26, *517*.

Statkewitsch (1904). Zeit. f. Allg. Physiol. 5, *511*.

*van Trigt, H. (1919). *A contribution to the Physiology of the freshwater Sponges.* Leiden.

Torrey, H. B. (1904). Biol. Bull. 6, *203*.

Valentin, G. (1842). Wagner's Handworterb. der Physiol. 1, *484*.

Vignon, P. (1901). Arch. de Zool. Exp. et gén., Sér. 3, t. 9, *371*.

*Williams, L. W. (1907). Amer. Naturalist, 41, *545*.

Yonge, C. M. (1926). Journ. Mar. Biol. Assoc. 14, *295*.

Chapter III

CILIARY CURRENTS

Methods of observing ciliary currents.

The most convenient method of observing the velocity and direction of ciliary currents is to note the movement of small particles of carmine or Indian ink placed in the neighbourhood of the ciliated surface. For such work the ciliated epithelium from the roof of the frog's mouth provides suitable material, but a much more striking and interesting structure is provided by the gills of Filibranch molluscs, of which *Mytilus* is a convenient example. The animal is opened by cutting through the adductor muscles and all the organs are removed except the gills and the under-lying mantle. One of the shells with its half of the mantle and its two gills is thoroughly washed in sea-water and laid submerged in a small dish of the same fluid. The ciliary currents (described in detail on p. 143) can readily be noted by suspending small particles of carmine in the sea-water by means of a pipette. The direction of the moving particles can be seen by the naked eye, and their rate determined by noting with a stop-watch the time required for movement between two fixed points on the surface of the tissue. The rate of movement depends very greatly on the temperature, and the following table indicates the average rate at which a small platinum weight is moved over the surface of the gills.

TABLE III.

Temperature ° C.	Speed of movement in millimetres per sec.
5	0·08
10	0·15
15	0·26
20	0·40

These figures shew that the rate of flow over quite an active ciliated epithelium approximates at room temperatures to $1\frac{1}{2}$ metres per hour. Considerably higher velocities are recorded by Copeland (1919) for the rate of movement of particles over the ciliated foot of the snail *Alectrion obsoleta*. In this case the speed was approximately 2 mm. per sec., or 7·2 metres an hour. This rate is the same as that at which the animal itself moves by ciliary activity. The temperature of the experiments is not recorded.

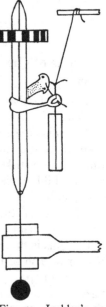

Various types of instruments have been devised for recording the rate of ciliary movement, but it is doubtful whether they give as good results as those obtained by direct observations. Apart from the difficulty of designing a very small revolving drum with reliable bearings, all ciliated epithelia are liable to contamination by mucus, which may materially affect the movement over localised areas. The movement of particles can be restricted to a small area of tissue and precautions taken to ensure that this area is free from mucus. For qualitative demonstration of ciliary movement the instrument designed by Inchley (1921) is both efficient and striking (Fig. 53).

Fig. 53. Inchley's apparatus for recording the activity of a ciliated epithelium.

A glass tube, 10 cm. long and about 5 to 8 mm. in diameter, is tapered down and sealed at one end, the other tapered end being left open and large enough to admit a vertically held hatpin which supports it. A strip of frog's ciliated membrane is looped round this, the ciliated surface being in contact with it; the glass spindle is thus slowly rotated by the activity of the cilia. Near its upper end the spindle carries a disc of cork about 3 cm. in diameter and graduated at the circumference to render the rotation more obvious. The loop of membrane is held together at its extremities upon a suitably bent pin, heavily weighted, and supported by a thread. The tension on the membrane is regulated by altering the inclination of this thread by movement of its fixed support. A drip feed of oxygenated Ringer solution is arranged so as to play on the glass spindle. For uniform motion the glass and cork must be well balanced on the needle; this can be attained by holding the needle inclined at 45° and rubbing off the heavier side of the cork until the spindle will stop dead in any position. Any slowing of the movement during the first hour is probably due to ropes of mucus tying the spindle, and these can be removed by means of a camel's hair brush.

More elaborate instruments embodying the same principle were designed by Engelmann (1877), and in these the speed of rotation of the cylinder was recorded by the frequency with which it made and broke an electrical circuit. So far nobody has succeeded in obtaining from these instruments results which cannot be obtained by direct observation, although they are of considerable interest from the point of view of technique or for the purposes of demonstration.

The velocity and pressure of ciliary currents.

That a ciliated epithelium can produce a current of water may readily be understood, but it is by no means easy to analyse the phenomenon in any detail. We may imagine that the practical effect of the ciliary stroke is to exert a force on a sheet of water lying in the plane of the beat, thereby causing the water to move over the surface of the epithelium. To do this the cilium must overcome the viscous resistance of the stationary water on each side. An analogy is provided by imagining the water to be composed of a series of layers like a pack of cards, and by representing the cilium as a needle. By pressing the side of the needle against the edge of one card the latter can be pushed out of the pack, and the force required is that which is necessary to overcome the resistance due to the friction exerted by the cards above and below the one which is moving. In the case of water currents there is, of course, a current behind the cilium as well as in front, whereby a continuity of movement is maintained.[1]

The course of water circulation set up by directional ciliary movement is illustrated by Fig. 54, which shews the currents set

[1] For an attempt to analyse ciliary currents by the behaviour of mechanical models, the article by Metzner (1920) may be consulted.

up by the peristomial cilia of *Stentor*, whilst Fig. 56 (from Lapage, 1925) shews the currents set up by the flagellum of *Codosiga botrytis*.

A clear picture of the mechanical results of flagellar movement is provided by the experiments of Parker (1914) on the sponge, *Stylotella*. By the activity of the flagellated chambers a current of water is drawn in at each ostium, and is expelled from the animal

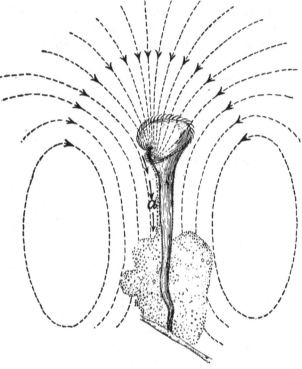

Fig. 54. Diagram of the currents caused by the peristomial cilia of *Stentor roeselii*. (After Jennings.)

by the exhalant oscula. The volume of water issuing from an osculum was determined by tying a glass tube of known diameter into the osculum of a sponge and submerging both sponge and tube in an aquarium. The velocity of flow through the tube was estimated by measuring the rate at which small particles of carmine were carried up the tube. It was found that the velocity of flow through a tube 17 mm. in diameter was approximately 4 mm. per second, so that 0·9 c.c. of water passed through the osculum every second. In *Leucandra aspersa* Bidder (1923) estimated the normal oscular velocity at 8·5 cm. per sec. An average sponge

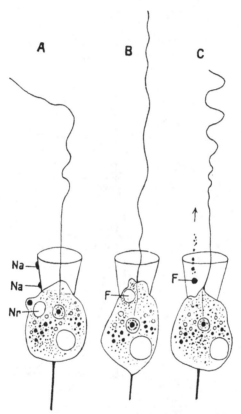

Fig. 55. *Codosiga botrytis.* Note the irregularity of the waves seen in the flagellum. (From Doflein.)

(*Stylotella*) with twenty oscula would pump more than 45 gallons of water through its body every day. On the other hand, the pressure of this current is very low and is equivalent to not more than 4 mm. of water. This value was determined by means of the apparatus shewn in Fig. 57. The force exerted by the sponge flagella is thus relatively small, and the whole of the energy imparted to the water is kinetic. It must be remembered, however, that the movements of the sponge flagella are uncoordinated (Bidder, 1923), and "the remarkable achievement of the perfected

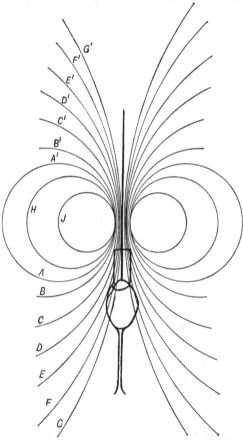

Fig. 56. Diagrammatic optical section of the currents created by the rotatory movement of the flagellum of *Codosiga*. The currents *D–G* are feeding currents; the more rapid currents, *A–F*, perform a cleansing function. (After Lapage.)

hydraulic organ in sponges is that from this waving of hairs $\frac{1}{100000}$ of an inch in thickness at a mean speed of seven feet an hour there is produced an oscular jet with an axial velocity of over half a foot a second (280 times the speed of the flagellum), which in

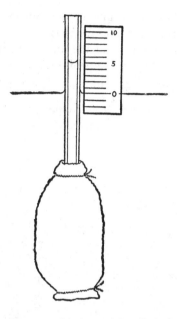

Fig. 57. Apparatus for determining the hydrostatic
pressure inside a sponge. (After Parker.)

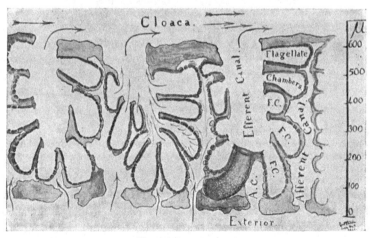

Fig. 58. Diagram of the wall of the sponge *Leucandra aspersa* shewing
the direction of the ciliary currents. (After Bidder.)

Leucandra throws to a distance of nine inches five gallons a day or a ton in six weeks" (Bidder, p. 303).

It is of interest to note that the kinetic energy of each c.c. of water issuing from the osculum of *Leucandra* is equivalent to a hydrostatic pressure of ·4 mm., since $V = \sqrt{2gh}$,

$$h = \frac{V^2}{2g} = \cdot 4 \text{ (approx.)},$$

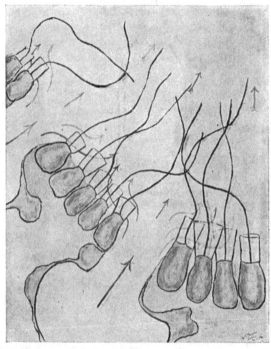

Fig. 59. Wall of a flagellated chamber of *Leucandra* shewing two afferent pores. The arrows indicate the direction of the currents set up by the uncoordinated movement of the flagella. (After Bidder.)

where h is the hydrostatic pressure, and V is the equivalent velocity. As Bidder points out, the values of 2–4 mm. recorded by Parker for *Stylotella* represent the pressure in the flagellated chambers. Before the water issues from the osculum a considerable percentage of this energy is lost owing to friction in the canals of the sponge. When we compare these pressures with those which are set up by muscular activity, it is obvious that ciliary activity would be quite inadequate for providing the blood of a large animal with sufficient energy to maintain an active circulation.

It will have been noted that the velocity with which water is moved over a flat ciliated surface is very much less than that at which the current leaves a narrow orifice such as the osculum of a sponge or the exhalent siphon of a mollusc. High velocities of ciliary currents are produced by directing the flow from an extensive ciliated surface into a large cavity with a narrow orifice. In this way the energy imparted to the water by the cilia is only subject to a minimum loss from frictional sources before emerging from the orifice (see Bidder, 1923).

The factors controlling the speed and frequency of the beat.

(a) *Size*. As a very rough but convenient rule the rate of beat of a cilium is inversely proportional to its size. Thus long flagella almost always beat less frequently than cilia, and large cilia usually beat more slowly than smaller types. The following table gives some idea of the number of beats carried out in one minute by different types of flagella and cilia, although, of course, they cannot be regarded as the result of strictly comparable observations.

TABLE IV.

Organism	Nature of motile organ	No. of beats per minute
Polytoma uvella	Flagellum	29
Oikommas vic. termo	,,	14
Euglena viridis	,,	67
Monas	{ Large flagellum	78
	{ Small ,,	94
Noctiluca viridis	—	5
Sponges	Choanocyte flagellum	min. 600
		max. 900–1200
Frog's oesophagus	Cilia	720 approx.
Gills of *Mytilus*	Frontal cilia	720 ,,

The speed of any rhythmical movement obviously depends on two factors: (a) the length of time occupied in moving through

one complete cycle, and (*b*) the time which elapses between the
end of one cycle and the beginning of the next; the second period
may be termed the 'latent' period. In the case of automatically
active cilia such as those of the frog or of molluscan gills the latent
period is extremely short, and observable alterations of activity
are nearly always due to changes in the time required for the
complete contractile cycle. A diagrammatic illustration of this is

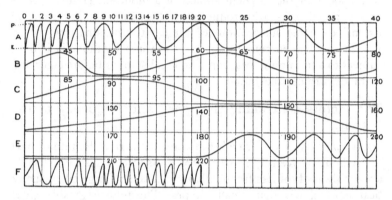

Fig. 60. Graphical representation of the effect of hydrogen-ions on the activity
of a frontal cilium of *Mytilus edulis*. The graph is composed of five sections; the
left side of each section is a continuation of the right side of the section immediately above it. The ordinates represent the amplitude of the ciliary stroke;
the base line represents the position of the cilium at the end of the effective
stroke. The abscissae are marked into 220 intervals of approximately $\frac{1}{2}$ sec. each.
The observations were begun soon after the cilium was exposed to air containing
CO_2; note that the period occupied by both phases of the beat gradually increased
until at about the end of *D* (160) the cilium ceased to beat. The CO_2 was then
turned off, and was replaced by dilute NH_3. Note that the recovery in speed
was very rapid. Throughout the whole experiment the amplitude of the beat
remained unchanged.

given by Fig. 60, which records the time occupied by the individual
beats of the frontal cilia of *Mytilus* when these were slowing down
by exposure to a current of CO_2, and when they were speeding up
again on removal of the gas. It will be noticed that all phases of
the beat are affected together.

In nearly all cases the effective stroke of a cilium is carried
out more rapidly than the recovery stroke; Kraft (1890) estimated
that in frogs' cilia the effective beat required only one-fifth the

time occupied by the recovery stroke, but it is doubtful whether this rule holds good in Ctenophores.

(*b*) *Temperature.* During normal life, temperature is by far the

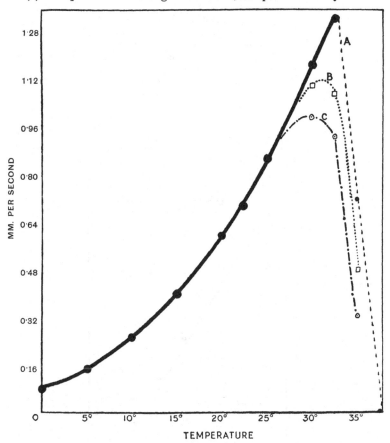

Fig. 61. Graph shewing the effect of temperature on the rate of movement of small particles by the frontal cilia of *Mytilus*. Curve *A* from observations made as soon as the tissue had reached the required temperature; curve *B* after 30 minutes; curve *C* after 1 hour. (After Gray.)

most important factor which influences the activity of cilia or flagella. Qualitative observations were made many years ago by Calliburces (1858) and other authors; in order to obtain quantitative values it is desirable to determine the temperature of the tissue by means of a small thermo-couple (Gray, 1923). Fig. 61 and Table V illustrate the quantitative effect of temperature on the frontal cilia of *Mytilus*.

TABLE V.

Temperature ° C.	Speed in mm. per sec.
0	0·08
5	0·15
10	0·26
15	0·40
20	0·60
22·5	0·70
25	0·86
30	1·17
32·5	1·33

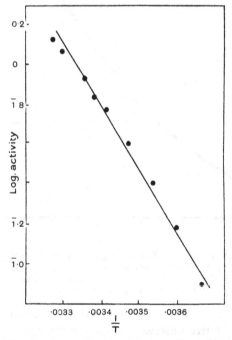

Fig. 62. Graph shewing the application of Arrhenius' empirical formula to the activity of the frontal cells of *Mytilus*. The abscissa represents the reciprocal of the absolute temperature.

Between 0° and 28° C. the effects of varying temperature are completely reversible; any rise in temperature rapidly involves an increase in the speed of the beat, but if the temperature is then lowered to its original value the speed also falls to its original value. The effect of temperature on the speed is reasonably accurately expressed by means of Arrhenius' equation

$$k_2 = k_1 e^{\frac{\mu}{2}\left(\frac{1}{T_1} - \frac{1}{T_2}\right)},$$

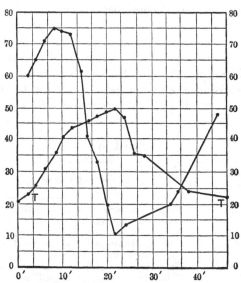

Fig. 63. Curves shewing the relation between temperature (curve TT) and rapidity of movement of the oesophageal cilia of the frog. The abscissae give, in minutes, the time from the beginning of the experiment. The ordinates give, for the temperature curves, the degrees Centigrade, and for the other curve the corresponding number of units of motor activity for the immediately preceding 2 minutes as registered by Engelmann's apparatus. Note that as the temperature rises beyond a certain point the activity falls, but rises again when the temperature is lowered. (From Davenport.)

where k_1 is the activity of the cilia at $T_1°$ (absolute temperature), k_2 is the activity at $T_2°$, and μ is a constant.

Above 28° C., however, a secondary effect of temperature can be observed. Individual cells begin to break away from the epithelium so that its apparent activity falls, and when the temperature is lowered the original activity is found to have been much reduced. This secondary effect is shewn in Fig. 61, in which the points on curves B and C represent observations made $\frac{1}{2}$ hour and 1 hour after the temperature had been raised, whereas curve A represents observations made as soon as the tissue had reached the desired temperature; that is, before disintegration had begun.

This secondary effect of high temperatures has no specific effect on the activity of the individual cells, but from about 34° C. onwards rising temperature no longer increases ciliary activity, although the rate of beat continues to increase. The reason for this is that at about 33° C. the cilium no longer passes through its full amplitude, so that although the frequency of the beat is extremely high, the mechanical effect is small. At about 37·5° C. the speed of the beat begins to fall, although there is no increase in amplitude. When the cilia come to rest at 40° C. they are in a state of heat coma; at 45° C. they pass into the contracted state of heat rigor.

TABLE VI.

Temperature ° C.	Direct effect on cilia of *Mytilus*
0–32·5	Progressive increase in speed with increasing temperature. Normal amplitude
34	Very rapid beat; amplitude reduced
36	Very rapid beat; amplitude very small, causing rapid flickering movement
37·5–38·5	Speed of beat rapidly reduced
40	Cilia stationary in relaxed position
45	Cilia pass into contracted position
47	Cilia opaque: irreversibly injured

All these phases are completely reversible but at 47° C. the cilia become coagulated and are killed. The depressant effect of heat is also illustrated by Engelmann's (1877) experiments with the ciliated oesophagus of the frog (Fig. 63), although in this case the maximum speed was not reached until about 44° C. and the cells were not killed even by exposure to 50° C.[1]

[1] According to Chamil (see *Tabulae Biol.* IV) maximum activity of the frog's cilia occurs at 35° C., whilst heat rigor occurs at 45°. These figures are almost identical with those found for *Mytilus*, and are considerably lower than those given by Engelmann.

The parallel between these facts and those which relate to muscular movement is striking. If the temperature of a frog's heart is raised, the speed of the beat increases up to 30° C. and the effect of any given range of temperature is almost exactly the same as that observed for cilia. Beyond 30° C., however, the amplitude of the beat falls markedly and eventually gives rise to a fibrillar condition just before the temperature at which the heart comes to rest. In the case of skeletal muscle, the height of an isotonic contraction rises with increasing temperature up to about 30° C.; at 32° C. it falls very rapidly until at 36° only a very slight response is made to stimulation; at 38° C. the highly contracted state of heat rigor sets in. To what extent these facts indicate a mechanism common to muscular and ciliated cells will be discussed elsewhere.

(c) *Hydrogen-ions*. The marked effect of hydrogen-ions on the rate of ciliary beat can only be discussed adequately in conjunction with the effect of other ions, but it is convenient to deal here with certain aspects of the problem for two reasons. First, variations of hydrogen-ion concentration in the cell have a much more powerful action on the speed of the beat than do variations in the concentration of any other ion. Secondly, variations in the concentrations of hydrogen-ions actually occur in nature, whereas other ions remain almost constant. The qualitative effect of hydrogen-ions in the external medium is shewn in Table VII, and quantitative data are given in Fig. 64 and in Table VIII.

All acids, when added to the external medium, are not equally efficient in inhibiting ciliary movement (Gray, 1922). Those acids which penetrate rapidly into the cell are more efficient than those which penetrate less readily, and a notable example of the second class is carbonic acid. It is well known that many bivalve molluscs are capable of living for many hours with their valves tightly closed: if a specimen of *Mytilus* is removed from the sea-water and left for two or three hours, it can be shewn that the concentration of CO_2 in the medium bathing the mantles and gills is sufficient to inhibit all ciliary movement. On washing the tissues in clean sea-water active movement is soon resumed. It is more than likely that the period during which *Mytilus* is normally uncovered at low tide is sufficient to produce enough CO_2 to inhibit

TABLE VII.

50 c.c. sea-water + c.c. N/10 HCl	pH	Activity of frontal cilia of *Mytilus* after								
		5′	10′	20′	30′	40′	60′	90′	120′	16 hrs
0·14	7·34	+ +	+ +	+ +	+ +	+ +	+ +	+ +	+ +	+ +
0·28	6·9	+ +	+ +	+ +	+ +	+ ⊕	+ +	+ +	+ +	+ +
0·42	6·7	+ +	+ +	+ ⊕	+	⊕	+ ⊕	+ ⊕	+ ⊕	+ +
0·56	6·6	+ +	+ ⊕	⊕	⊕	Very slight	Very slight	⊕	+ ⊕	+ +
0·70	6·2	+ +	+ ⊕	⊕	⊕	Very slight	Very slight	Very slight	Slight	+ +
0·98	5·5	+	⊕	⊕	o	o	o	o	o	o
1·19	4·2	Very slight	o	o	o	o	o	o	o	o
1·33	3·8	o	o	o	o	o	o	o	o	o
1·40	3·6	o	o	o	o	o	o	o	o	o

 + + Very active.   + ⊕ Active.   + Considerable movement.   ⊕ Slight.

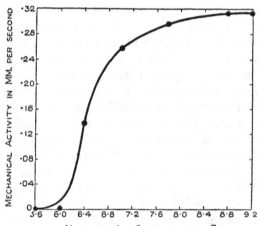

Fig. 64. Graph shewing the effect of hydrogen-ions on the mechanical activity of the frontal cilia of *Mytilus*. (After Gray.)

TABLE VIII.

Rate of movement in mins. per sec.	pH
o	5·6
o	6·0
0·14	6·4
0·26	7·0
0·30	7·8
0·31	8·8
0·31	9·2

its cilia, and thereby reduce the O_2 requirements of the animals (see p. 104). Haywood (1925) has shewn experimentally that the cilia of *Mytilus* are influenced both by the pH of the external medium and by its percentage saturation of CO_2.

An interesting example, illustrating the relationship between cilia and the acidity of their normal environment has been described by Yonge (1925).

This author found that the concentration of hydrogen-ions necessary to bring to rest the cilia of *Mya* depends on the concentration with which they are normally in equilibrium. (Table IX.)

TABLE IX.

Cilia from	Minimum pH in which they can function	Average pH of fluid normally round them
Style sac	3·5–4·0	4·45
Stomach	4·0	5·8
Mid-gut	4·4–4·8	6·2
Oesophagus	4·4–4·8	6·6
Rectum	4·4–4·8	6·9
Gill	5·2–5·8	7·2

(*d*) *Electric currents.* The movement of many cilia appears to be uninfluenced by the presence of an electric field, but according to Segerdahl (1922) the latero-frontal cilia on the gills of *Anodon* beat with increased rapidity if a current of one milliampere per square millimeter is passed from the base of the gill towards the tip. If the direction of the current is reversed the cilia are brought to rest. To some extent the nature of the response depends upon the strength of the current used.

The reactions of ciliated protozoa to electric fields have been discussed by Jennings (1915), and in certain cases the cilia nearest to the cathode reverse the direction of their effective stroke.

Conditions of flow in ciliated tubes.

When a current of water is being driven over the surface of a ciliated epithelium, it is of interest to know the extent to which the effects of the cilia are transmitted in a direction at right angles to the surface. If *AB* (Fig. 58) is a sheet of water moving horizontally over the surface of the epithelium, then the layer (*CD*) of water immediately above *AB* tends to move forward with AB owing to the cohesive properties of the water, and at the same time DC tends to retard the flow of AB. In this way the effect of the cilia is transmitted for some distance at right angles to the effective movement, but the further away from the driving surface the lower will be the velocity imparted to the successive layers of water. If at any moment a layer of particles (AW) is arranged vertically over the epithelium then, after a unit time, they will be distributed along the curve XB.

It is clear that at some distance away from the ciliated surface the rate of flow will be negligible. The rate of change of velocity is pro-

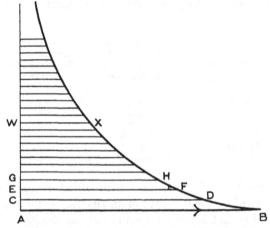

Fig. 65. Diagram to illustrate the effect of a ciliated surface, upon the movement of superimposed layers of water. If the length AB represents the velocity of flow of the water immediately above the cilia, then the velocity of flow at any level such as at $C, E, G \ldots$ or W is given by the points on the curve $D, F, H \ldots X$.

portional to the distance any particular layer is situated above AB, and inversely proportional to the coefficient of viscosity. From observation of actual surfaces it looks as though the effect of cilia becomes very small beyond a distance equal to four or five times the length of the cilia. This principle is of importance in estimating the efficiency of cilia when these are used as a motive power for driving fluid through tubular organs.

In all ciliated tubes the velocity of flow due to either of the walls must decrease parabolically as the distance from the ciliated wall. As long as the radius of the tube does not exceed four or five times the length of the cilia we may expect a reasonably uniform flow to be maintained across the whole section, since the speed imparted to the centre of the fluid by both ciliated walls is not much lower than that at the walls themselves; in wide tubes, however, the fluid at the centre is almost stationary.

In narrow ducts such as the nephridial tubes of an earthworm or the *vasa efferentia* of the vertebrate testis the only means of propulsion are the cilia, and these are quite efficient. In the gut, however, where the tube is wide, the fluid must be moved by muscular contraction, for in this case efficient ciliary circulation would be limited to a very small fraction of the whole cross-section. So also in the blood system cilia would be quite ineffective except in the small veins of an animal content to have a very sluggish circulation.

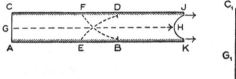 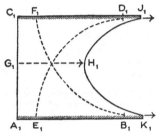

Fig. 66. Diagrammatic illustration of the effect of diameter on the velocity of flow through ciliated tubes. In both tubes the velocity imparted to the driving surface is assumed to be equal and represented by (AB, CD, A_1B_1, C_1D_1). The velocity imparted by each driving surface across the section of the tubes is shewn by BF, DE, B_1F_1, D_1E_1. The velocity in the narrow tube, GH, is seen to be greater than in the wider tube, G_1H_1, although the power per square centimetre of surface is the same in both.

It will be realised that with any particular type of cilium the narrower the tube the more energy is imparted to each unit mass of water. In a circular tube 1 cm. in length the total driving force of the cilia is $2\pi rP$, where r is the radius of the tube and P is the force exerted by 1 sq. cm. of ciliated surface. The mass of water in the tube is πr^2, so that the resulting acceleration is $\dfrac{2P}{R}$: in other words the average amount of energy imparted to 1 c.c. of water in unit time is inversely proportional to the radius of the tube.

For a description of ciliary currents occurring in tubular organs, see Chapter VIII.

References

Bidder, G. P. (1923). Quart. Journ. Micro. Sci. 67, *293*.

Calliburces, P. (1858). Comptes Rendus, 47, *638*.

Copeland, M. (1919). Biol. Bull. 37, *126*.

Engelmann, T. W. (1877). Pflüger's Archiv, 15, *493*.

Gray, J. (1920). Quart. Journ. Micro. Sci. 64, *345*.

—— (1922). Proc. Roy. Soc. 93 B, *104*.

—— (1923). Proc. Roy. Soc. 95 B, *6*.

—— (1924). Proc. Roy. Soc. 96 B, *95*.

Haywood, C. (1925). Journ. Gen. Physiol. 7, *693*.

Inchley, O. (1921). Proc. Physiol. Soc. *1*.

Jennings, H. S. (1915). *Behaviour of the Lower Organisms*. New York.

Kraft, H. (1890). Pflüger's Archiv, 15, *493*.

Lapage, G. (1925). Quart. Journ. Micro. Sci. 69, *471*.

Metzner, P. (1920). Biol. Zentralb. 40, *49*.

Parker, G. H. (1918). *The Elementary Nervous System*. Philadelphia.

Segerdahl, E. (1922). Skand. Arch. Physiol. 42, *62*.

Yonge, C. M. (1925). Journ. Mar. Biol. Assoc. 13, *938*.

Chapter IV

THE FORCE EXERTED AND THE WORK DONE BY CILIA

The first attempt to estimate the force exerted by co-ordinated ciliary effort was made by Engelmann (1877), who concluded that the oesophageal epithelium of the frog could exert a pressure of approximately 0·4 gram per square centimetre. A more complete series of experiments with the same material was performed by Maxwell (1905). The method used for determining the work done was essentially the same as that previously employed by Bowditch (1876); the epithelium was fixed to the surface of an inclined plane and the rate observed at which loads of varying mass, but of constant surface area, moved up a damp ciliated surface. The results of a typical experiment are recorded below.

TABLE X.

Weight of load in grams (M)	0·415	1·66	2·905	4·15	5·395	6·64	7·885	9·13
No. of secs. required by 58 sq. mm. to raise the load through 0·44 mm.	9	15	19	24	30	48	310	∞
Velocity of movement (V) = mm. moved per min.	2·93	1·76	1·39	1·10	0·88	0·55	0·085	0
Work done per min. per sq. cm. of surface $W = M \times V$	2·117	5·08	7·018	7·937	8·255	6·35	1·167	0

Maxwell concluded from these figures that the rate at which cilia perform work increases with increasing load until a maximum is reached, after which the rate rapidly declines (see Fig. 67). These results can be interpreted to mean that with increasing load the rate at which the cells expend energy is increased, and that the failure at higher loads is associated with fatigue. This was in fact the view adopted by Maxwell.

It is obvious, however, that this conclusion is based on the assumption that the whole of the work done by the cilia is performed against gravity and entirely ignores the work done against the resistance of the water lying between the driving surface and the load. In Maxwell's experiments the total work done by the cilia is the sum of the work done in raising the load through a height h and the work done against the force of water resistance (f) through a distance $h \cos \alpha$, where α is the angle of inclination

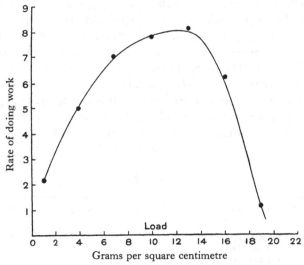

Fig. 67. Efficiency of the oesophageal epithelium of the frog under varying loads. The ordinates are gram-millimetres of work per minute. Compiled from Maxwell's data.

of the plane. The actual value of $h \cos \alpha$ in Maxwell's experiments was $10h$, and this will be represented in the following analysis by V, as it is proportional to the surface velocity.

When a ship is moving fairly rapidly through the water the resistance (f) of the water varies with the square of the velocity, but the very slow speeds involved in ciliary movement exempt it from this law owing to the absence of eddies. For slow speeds the resistance of the water becomes proportional to the velocity and not to its square.

The work done in unit time at a velocity V is therefore $kV \cdot V$, where k is a constant varying with the coefficient of viscosity of the medium. The total work done by the cilia per unit time is therefore given by the equation

$$W \quad = \quad Mh \quad + \quad kV^2.$$

total work   work done   work done
against gravity   against viscosity

Table XI is based on the average velocity of the loads in ten of Maxwell's experiments; if k is given the value ·007, the total work done on loads varying from ·415g to 5·395g is approximately constant at 10g mm. per minute. When moving along a horizontal plane all the loads ought to travel with the same velocity, which can be calculated $\left(V = \sqrt{\dfrac{W}{k}} \right)$, since the work done against gravity is zero. The calculated value of 38 mm. per minute agrees reasonably well with observed velocities.

TABLE XI.

Load in grams	0·415	1·66	2·905	4·15	5·395	6·64
Average time in secs. to move through a vertical distance of 0·44 mm. (Driving surface 58 sq. mm.)	8	12	15	20	26	36
Vertical vel. in mm. per min.	3·3	2·2	1·76	1·32	1·01	0·73
Work done in g mm. per min. against gravity per 100 sq. mm.	2·36	6·30	8·80	9·44	9·30	8·40
Linear velocity (V) in mm. per min.	33	22	17·6	13·2	10·1	7·3
Work done against friction kV^2 [k = 0·007]	7·7	3·39	2·17	1·22	0·71	0·37
Total work done per min.	10·06	9·69	10·97	10·56	10·01	8·77

It is obviously desirable to check these conclusions by observing the velocities imparted to a constant load under different rates of ciliary beat. At present the only known means of varying in a regular manner the speed of horizontal movement is provided by changes in temperature, and the only way of estimating the total expenditure of energy (W) is based on the assumption that the rate of oxygen consumption is a measure of the output of energy.

Fig. 68 shews that when the temperature varies, the oxygen consumption of *Mytilus* gills varies directly with the speed of the ciliary current. At first sight this looks as though the ratio of

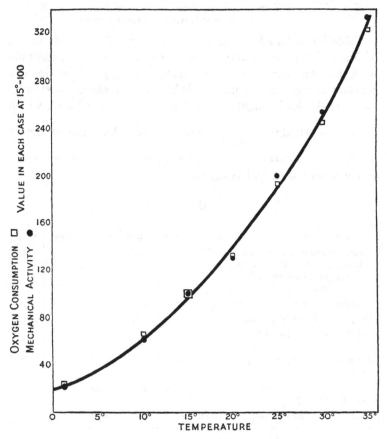

Fig. 68. Graph shewing the effect of temperature on the mechanical activity and on the rate of respiration of *Mytilus* gills. (After Gray.)

work done to the energy expended rises with increasing temperature. But it is necessary to remember that with increasing temperature a fall occurs in the coefficient of viscosity.

In Table XII the values obtained experimentally are corrected for changes in viscosity and the results plotted in Fig. 70.

TABLE XII.

Temperature $T°$	Velocity V	Coefficient of viscosity η	Rate of doing work $V^2\eta$	Rate of O_2 consumption O_2
10	6	0·0130	0·47	65
15	10	0·0110	1·10	100
20	13	0·0100	1·69	130
25	19	0·0085	3·06	184
30	24	0·0080	4·61	246

There appears to be reasonably good agreement between the experimental data and the demands of general theory, but the validity of these results must depend on the results of further work. It may be mentioned, however, that the direct relationship between the speed of the ciliary current and the rate of oxygen consumption finds a parallel in cardiac muscle (Evans, 1912).

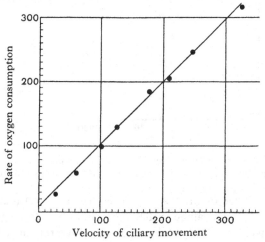

Fig. 69. Graph shewing the relationship between the velocity of movement of a small platinum plate over the surface of *Mytilus* gill and the rate of oxygen consumption of the tissue. (From Gray.)

Attempts have also been made to estimate the force exerted by cilia by taking advantage of the fact that certain ciliates are negatively geotropic, so that when exposed to centrifugal force the cilia tend to oppose the force. By adjusting the centrifugal force to the value at which the ciliates are just unable to orientate themselves normally, Jensen (1893) concluded that the ciliated surface of one *Paramecium* could exert a force of ·00017 milligram, or roughly nine times its own weight; each cilium exerts a force of approximately $4·5 \times 10^{-7}$ mg., and each square centimetre of surface a force of 21 mg. At the same time Jenson estimated that 89 per cent. of the available energy is used up in overcoming the

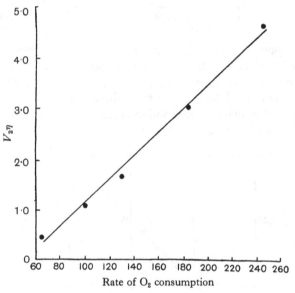

Fig. 70. Figure shewing relationship between the amount of work done by the frontal cilia of *Mytilus* against the viscosity of water, and the rate of O_2 consumption.

viscous resistance of the water. An approximate estimate of the efficiency of the cilia alone is that they can exert a force 368 times their own weight. From evidence based on the rate of sedimentation of motile bacteria Angerer (1919) concluded that each organism could perform work at rates varying from $1,2 \times 10^{-14} - 26 \times 10^{-13}$ gram centimetres per second.

It is interesting to compare the work done by a ciliated cell with that by a muscle. Hartree and Hill (1920) record that the work done by a frog's muscle at 21° C. is equivalent to raising the muscle through 6 metres in 0·4 sec., which is equivalent to

900 metres a minute; according to Engelmann a ciliated cell can only raise itself 4·25 metres. Admitting that the latter estimate is probably too low, it is still quite evident that the horse-power of muscle is of a much higher order than that of a cilium. The only advantage gained by ciliary movement as opposed to muscular movement is that the former can be very prolonged whereas a muscle working at a much higher pitch is liable to fatigue.

Efficiency of Cilia.

It is obviously impossible from the data available to obtain any real insight into the efficiency of cilia as propulsive units. The whole subject is beset with complex hydrodynamical problems which are, as yet, unsolved. We may, however, consider two factors of biological importance: (i) the ease with which a ciliated animal, starting from rest can attain its maximum velocity; (ii) the ease with which an animal in motion can move in any desired direction. Let us assume that a *Volvox* weighing M grams is swimming in a horizontal direction at its maximum speed (V centimetres per second); by Stokes' Law the viscous resistance is directly proportional to the velocity and is $6\pi r\eta v$, where r is the radius of the *Volvox* and η is the coefficient of viscosity of water. If the cilia suddenly cease to beat it can be shewn that the kinetic energy of the organism will only carry it forward against the resistance of the water for an extremely short distance.

$$\text{Acceleration} = \frac{\text{Retarding force}}{\text{Mass}},$$

$$\frac{\partial v}{\partial t} = \frac{-kv}{M}, \text{ where } k = 6\pi r\eta,$$

$$v = Ve^{\frac{-kt}{M}} \qquad \qquad \text{......(i).}$$

The distance travelled before coming to rest is $\int_0^\infty v.\partial t,$

$$\int_0^\infty v.\partial t = \int_0^\infty Ve^{\frac{-kt}{M}}.\partial t = \frac{VM}{k}.$$

Let the radius (r) of the *Volvox* be ·05 cm.
Let the density (ρ) of the *Volvox* be 1·01.
Let the initial velocity (V) of the *Volvox* be ·1 cm. per sec.

Let the temperature be 15° C., giving a value for η of ·011.

$$\frac{VM}{k} = \frac{\frac{4}{3}\pi r^3\,(\rho)\,V}{6\pi r\eta} = \frac{2r^2\,(\rho)\,V}{9\eta} = \text{·005 cm.}$$

$$= 50\,\mu\ (= \text{one-tenth of its own length}).$$

It follows from (i) that this very short distance, viz. $50\,\mu$, will only be reached after an infinitely long time, so that we can safely assume that from all practical points of view a small ciliated organism comes to a dead stop as soon as it switches off the motive power of its cilia. Correspondingly it starts off again at full speed as soon as the cilia become active. From the point of view of gliding through the water with stationary cilia the kinetic energy of small aquatic organisms is hopelessly inadequate since the viscous resistance of the water is of a relatively high order.

If, however, we take a larger organism, *e.g.* 1 cm. in diameter, moving with a velocity of 1 cm. a second, then with stationary cilia it will glide a distance $\left(\dfrac{VM}{k}\right)$ of 20 cm. Here we have quite a different state of affairs, and the animal may have grave difficulties in avoiding obstacles, and when starting from rest it will only reach its maximum velocity after some considerable time. A few ciliates, notably *Cyclidium glaucoma*, are reported to acquire an amount of kinetic energy sufficient to overcome the viscous resistance of the water for an appreciable distance. *Cyclidium* is a holotrichous ciliate which often lies quite still with all the cilia rigid and outspread. "After a short interval, but long enough for observation, the animal disappears with a sudden jump, to be found again a short distance away. The sudden motion appears to be effected by the terminal bristle" (de Morgan, 1926, p. 37). It would be interesting to know whether the cilia, other than the terminal bristle, are in motion during the 'jump.'

In general, however, as an aquatic animal increases in size, the need for methods of locomotion more powerful than those provided by cilia or flagella, becomes more and more acute. For heavy animals requiring a wide and powerful range of movement cilia are quite inadequate as organs of locomotion, and the greater horse-power developed by muscular units becomes of overwhelming importance.

Since the viscous resistance of the water increases with the velocity of the moving body, it follows that the maximum velocity of a ciliated organism is reached when the force exerted by the cilia equals the retarding force of viscosity. For a spherical animal (if P is the force in dynes developed by the cilia per square centimetre of surface),

$$4\pi r^2 P = 6\pi r \eta V,$$

$$V = \frac{2Pr}{3\eta}.$$

The viscous resistance varies with the radius, but the motive power varies with the surface so that as long as the limits of Stokes' Law are not exceeded, the larger the animal the higher is its maximum velocity. Hence with cilia of equal power, the relative speed of a young *Volvox* colony ought to be the same as that of an older and larger colony, but the absolute speeds will increase with growth of the organism. It also seems reasonable to assume that animals elongated in the axis of movement will be able to move faster than spherical animals of the same cross-section, and this may account for the fact that most actively moving ciliates approximate in form to the stream lines of a torpedo.

So far we have only considered motion in a horizontal plane, but an animal often requires to move upwards in the medium. It is clear that unless the density of the animal is exactly the same as that of the medium in which it lives, any upward motion against gravity must either be provided for by more energetic or frequent ciliary beats or it can only be carried out at reduced speed. Assuming that we are correct in regarding the total energy expended per second by a cilium as independent of the load, it is possible to consider the relationship of movement to gravity. If a Ctenophore of mass M moves vertically upwards with a given velocity (V), then the work done in unit time is equal to the sum of the work done against gravity and the work done against the resistance of the water,

$$W = M(\sigma - \rho)V + kV^2,$$

where σ = density of the organism and ρ = density of the medium. When moving in a horizontal plane all the energy is accounted for in the second term, so that the maximum velocity is raised from $\frac{1}{k}\sqrt{W - M(\sigma - \rho)V}$ to $\frac{1}{k}\sqrt{W}$ when movement changes from the vertical upward type to the horizontal. It is highly unlikely, however, that the work done against gravity can be very great compared to that done against the resistance of the water (Jensen gives the ratio of 1 : 10). Flagellates and Ctenophores can swim vertically upwards at speeds quite comparable to those in horizontal movements, and while they are rising there is no evidence that slight disturbances which rotate the animal through 90° lead either to rapid movement away from the original line of movement or to a marked reduction in the speed of the beat. Normally by far the larger percentage of the animal's energy is used up against the resistance of the water and very little against gravity.

From these considerations we may conclude that ciliary locomotion is suitable for small animals of low specific gravity which are content to move about at low speeds, and that the best results are obtained by animals which are elongated in the direction of movement. For very small animals whose surface is richly ciliated the specific gravity is relatively less important; but if the size increases the body must have a high water content. A few examples readily suggest themselves: e.g. *Paramecium*; the ciliated larvae of invertebrates which are characterised by a large blastocoelic cavity; and among larger animals the Ctenophores with their 'gelatinous' bodies of low specific gravity; all these animals are admirably adapted for ciliary locomotion and are free to move with almost equal facility in all directions.

References

v. Angerer, K. (1919). Arch. f. Hyg. 88, *139*.
Bidder, G. P. (1923). Quart. Journ. Micro. Sci. 67, *293*.
Bowditch, H. P. (1876). Boston Med. and Surg. Journ. 15, *159*.
Engelmann, T. W. (1877). Pflüger's Archiv, 15, *493*.
Evans, C. L. (1912). Journ. of Physiol. 45, *213*.
Hartree, W. and Hill, A. V. (1920). Journ. of Physiol. 54, *84*.
Jensen, P. (1893). Pflüger's Archiv, 54, *537*.
Maxwell, S. (1905). Amer. Journ. Physiol. 13, *154*.
de Morgan, W. (1926). Journ. Mar. Biol. Assoc. 14, *23*.

Chapter V

THE RELATION OF CILIARY MOVEMENT
TO IONIC EQUILIBRIA

Ionic equilibria.

The relationship between the behaviour of ciliated tissues and the ionic composition of their surrounding medium has attracted a certain amount of spasmodic attention for some years. A really adequate analysis of the problem has, however, been hampered by the fact that different workers have used different materials and have often overlooked or ignored the fact that it is not easy to interpret biological experiments which involve more than one variable. R. S. Lillie (1906) carried out an extensive series of experiments with excised fragments of the gills of *Mytilus*, and concluded that both anions and cations affect the rate of ciliary movement. A. G. Mayer (1910) worked with veliger and other larvae, and concluded that at least the cations influence ciliary movement in the reverse manner to their effect on muscular contraction. Before attempting to assess the value of the results obtained by these workers it is perhaps convenient to see how far the main problem can be analysed from first principles.

In the first place, it must not be forgotten that nearly all living tissues begin to shew signs of disintegration if there is a serious change in the concentration of certain of the ions in their environment. For example, all tissues without exception become unstable and die if entirely deprived of those elements which belong to the same chemical groups as magnesium and calcium. This fact can very readily be demonstrated in ciliated tissue. The normal external environment of the gills of *Mytilus* is sea-water, and this medium contains sodium, potassium, magnesium, calcium and hydrogen-ions (pH 8·1 (approx.)). In artificial sea-water of normal composition, excised gills remain alive and active for very many hours. If, however, the magnesium and the calcium are removed, then at ordinary temperatures the whole tissue begins to break up within a few hours. The most striking changes seen are the dissolution of the intercellular matrix and the absorption of water by the cells (Gray, 1922, 2; 1926, 2); the direct effect on the cilia is comparatively small, and isolated cells move actively through the medium by their own activity. It is obvious that considerable care must be exercised to avoid confusion between the effect of the presence or absence of a particular ion on the general stability of the cell and its direct effect on the ciliary apparatus. Since the former effect is irreversible it is safest to base conclusions concerning the second effect upon such experiments as are reversible.

The following analysis refers mainly to excised fragments of the gills of *Mytilus* and unless specifically stated to the contrary, the cilia observed were the *frontal* cilia at the distal ends of the gill-filaments.

(*a*) *Cations and anions.* If a series of artificial media are prepared, in which the concentrations of hydrogen-ions and all other cations are the same as in sea-water, but in which the chlorine-ions are entirely replaced by other anions, comparatively little change occurs either in the rate of ciliary movement or in the stability of the tissue (Gray, 1922, 2).

TABLE XIII.

The effect of mixtures of Na·, K·, Ca·· and Mg·· (proportions as in sea-water and pH 7·8) upon the movement of the frontal cilia of Mytilus.

Anions	Degree of movement after				
	30 mins.	1 hour	2 hours	3 hours	5 hours (+)
Chlorides	+ +	+ +	+ +	+ +	+ +
Nitrates	+ +	+ +	+ +	+ +	+ +
Iodides	+ +	+ +	+ +	+ +	+ +
Bromides	+ +	+ +	+ +	+ +	+ +
Acetates	+ +	+ +	+ +	+ +	+ +
Sulphates	+ +	+ +	+ +	+ +	+ +
Tartrates	+ +	+	⊕	o	o
Citrates	o	o	o	o	o

+ + = Normal activity   ⊕ = Slight activity
+ = Reduced activity   o = No activity

According to R. S. Lillie, however, the anions can be arranged in a definite order of inhibitory power

$$SCN > I > Br, NO_3 > SO_4 > Cl, \text{ acetate.}$$

There can be little doubt that this series is erroneous and that the facts observed by Lillie really illustrate the order in which the sodium salts with different anions influence the stability of the tissue when calcium and magnesium are absent. A comparison of Tables XIII and XIV makes this fairly clear (see also Gray, 1926, 2).

TABLE XIV.

The effect of pure isotonic sodium salt solutions pH 7·8 upon the epithelium of Mytilus gills.

Anion	Time			
	30 mins.	1 hour	2 hours 3 hours	5 hours
Chloride	Some cilia active	Cells considerably swollen. Some cilia destroyed	Tissue disorganised. Cells much swollen. Cilia destroyed	—
Nitrate	A few cilia active. Most cells begin to swell	Cells much swollen. Few cilia left	Tissue disorganised. Cells much swollen. Cilia destroyed	—
Bromide	A few cilia active. Most cells begin to swell	Cells much swollen. Few cilia left	Tissue disorganised. Cells much swollen. Cilia destroyed	—
Iodide	Nearly all cells markedly swollen	Tissue disorganised. Cells much swollen. Cilia destroyed	—	—
Acetate	No movement. Cells normal	No movement. Cells slightly swollen		
Sulphate	No movement. Cells do not swell. Cilia remain healthy in appearance			
Tartrate	No movement. Cells do not swell. Cilia remain healthy in appearance			
Citrate	No movement. Cells do not swell. Cilia remain healthy in appearance			

The only anions which directly affect ciliary movement to any appreciable extent are those which interfere with the normal balance of the cations present. Tartrates and citrates convert the calcium and magnesium to an unionised condition (see Gray, 1922, 2), and hydroxyl-ions reduce the concentration of hydrogenions. In subsequent paragraphs it will be seen that although ciliary movement is not visibly affected by anions it is definitely affected by cations. Since there is no evidence to shew that the cells contain colloids of the acid side of their isoelectric point, the indifference displayed to anions is not surprising. The great bulk of the cell contents are quite clearly on the alkaline side of their isoelectric point, so that the rôle played by cations on the ciliary apparatus is conceivably due to their general effect on the degree of aggregation and hydration of colloidal particles; at any rate, it is unnecessary and probably wrong to attribute the action of cations to a hypothetical power of entering the cell, and the indifference displayed to anions to a lack of penetrating power. For a further account of the action of pure sodium salts on ciliated tissue reference may be made to Gray (1926, 2).

(b) *Monovalent cations.* By far the most important ion, viz. that of hydrogen, has already been considered (Chapter III), but the monovalent metallic ions are also of some importance.

In sea-water the concentration of sodium is roughly 37 times as great as that of potassium, and it is easy to shew experimentally that considerable variations can be made in the concentration of sodium without markedly affecting ciliary activity. That the mechanism is not entirely unaffected by this metal is shewn by the fact that if the whole of the sodium is replaced by other monovalent cations the rate of ciliary beat is quite clearly and rapidly affected: the rate of beat is least in Li· and greatest in K·.

TABLE XV.

Solution	pH	Movement of terminal cilia					
		2 mins.	5 mins.	10 mins.	20 mins.	30 mins.	60 mins.
LiCl, MgCl$_2$, CaCl$_2$	7·0	o	o	o	o	o	o
NaCl, MgCl$_2$, CaCl$_2$	7·2	+ +	+ +	+ +	+ +	+ +	+ +
NH$_4$Cl, MgCl$_2$, CaCl$_2$	7·0	+ +⊕	+ +⊕	+ +⊕	+ +⊕	+ +⊕	+ +⊕
KCl, MgCl$_2$, CaCl$_2$	7·2	+ + +	+ + +	+ + +	+ + +	+ + +	+ + +
Sea-water (HCl)	7·2	+ +	+ +	+ +	+ +	+ +	+ +

Li· < Na· < NH·$_3$ < K·

o = No movement.   + +⊕ = Quicker than normal.
+ + = Normal rate of heat.   + + + = Very rapid beat.

Simultaneously with the variation in the rate of beat goes a corresponding variation in the rate at which the cells absorb oxygen (Gray, 1924). This series of monovalent ions is precisely similar to that observed if ciliated tissues are placed in solutions of the pure chloride solutions (Höber, 1914), but in the latter case the facts are greatly obscured by the disintegrative action of the solutions on the tissue.

As is so frequently the case with other tissues, there is a marked difference between sodium and potassium in their effects on ciliary movement. As far as the frontal cilia are concerned the difference is not particularly striking, apart from the fact that the cilia beat faster in the medium containing abundance of potassium than in the medium containing a lower concentration. Biologically it is interesting to find a process which will continue with only a change in velocity in concentrations of potassium varying from a very low value up to 0·5 molecular; but the behaviour of the lateral cilia illustrate the more fundamental properties of this metal.

The *lateral* cilia on the gills (see Fig. 99) readily come to rest when excised fragments are placed in any medium containing more magnesium than is present in the blood. Such a medium is obviously supplied by sea-water. In this respect the *lateral* cilia differ markedly from the *frontal* cilia which continue their activity for prolonged periods in sea-water. It is the peculiar property of potassium to antagonise this action of magnesium on the lateral cilia. Sodium has no such antagonistic power, and it is interesting to note that the same difference between sodium and potassium is observable in their action on the stability of the intercellular matrix. It has been shewn (Gray, 1922, 2) that the normal condition of stability is maintained by magnesium. A mixture of sodium and magnesium allows stability to persist: a mixture of potassium and magnesium leads rapidly to instability. It looks very much as though potassium could eject magnesium from certain positions in the cell and that this property is not shared by sodium.

On the slowly moving *latero-frontal* cilia of *Mytilus*, potassium has an effect not seen in the case of the other cilia. The *latero-frontal* cilia at first respond to the presence of potassium by an increase in speed without any change in the form of the beat. If, however, the concentration of potassium is high, then the recovery stroke becomes incomplete, and the cilia vibrate very rapidly with a much reduced amplitude (Fig. 71). The 'tone' of the cilium thus appears to be increased, although after a time the amplitude gradually increases and the normal beat is resumed.

(c) *Divalent ions*. The majority of divalent cations are comparatively uninteresting: if the calcium and magnesium in sea-water are replaced by other divalent metals no very striking changes can be observed in activity before the whole tissue begins to shew signs of irreversible change. The divalent ions of calcium and magnesium are, however, of interest, and must be considered separately.

To any increase in the concentration of calcium in the surrounding medium, ciliated cells are remarkably indifferent,

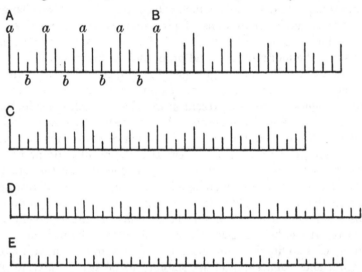

Fig. 71. Diagram illustrating the effect of excess of potassium on the beat of the fronto-lateral cilia. a = end of recovery beat. b = end of effective beat. A–B normal beat: B–C, 1st stage of contraction effect: C–D, 2nd stage: D and E, final stage in which cilia remain for a considerable period. The cilia are seen in the plane of their beat, *i.e.* at right angles to the plane of the paper.

although the secondary changes induced in the tissue are quite typical and well-marked (Gray, 1922, 1). As far as the actual mechanism of movement is concerned the excess of calcium involves little change. At first sight this is very surprising, since calcium has a well-marked and powerful action on the state of colloids on the alkaline side of their isoelectric point. Inside the cell, however, there is a comparatively high concentration of inorganic phosphates —sufficient to give a well-marked molybdate test; there is also a considerable concentration of bicarbonate ions. Now in any system containing $Ca^{..}$, $H^{.}$, CO_3'', HCO_3' and phosphate-ions, the number of calcium *ions* is fixed by the concentration of the other ions (Kugelmass and Shohl, 1924). Thus in a simple case

$$Ca^{..} = \sqrt{\frac{k\,(H^{.})}{(HCO_3')\,(HPO_4'')}}.$$

From this it appears that a temporary increase can occur in the concentration of calcium-ions outside the cell without causing any marked increase in these ions inside, although the cell may be freely permeable to them. An alternative explanation would be available if it could be shewn that the cells are not permeable to calcium-ions.

To an absence of calcium the cilia react by a gradual decrease in the rate of beat; eventually all movement ceases (Gray, 1924). The visible changes are precisely similar to those seen during inhibition by acid: all phases of the beat are slowed and the frontal cilia stop at the end of the effective beat. There is, however, one significant difference between the two phenomena. When ciliary movement ceases owing to the presence of hydrogen-ions there is a marked fall in the oxygen consumption; when movement ceases owing to lack of calcium the oxygen consumption remains at its normal level. The effect of $Ca^{..}$ lack is reversible, but the process is slow (Gray, 1924). The significance of these facts is discussed on p. 109.

Unlike calcium, magnesium plays an essential rôle in maintaining the normal stability of the intercellular matrix (Gray, 1926, 2). As long as the concentration of magnesium is adequate for this purpose great variations in its concentration may be made without inducing any marked change in the degree of activity of the *frontal* cilia: these cilia will beat for many hours in a solution of magnesium chloride isotonic with sea-water (approx. 0·4 mol. $MgCl_2$). On the other hand, this metal exerts a well-marked action on the *lateral* cilia. When a piece of excised gill is placed in normal sea-water the lateral cilia cease to move after a period which varies from 10 minutes to 2 hours. If the tissue be then transferred to any solution which does not contain magnesium the *lateral* cilia quickly shew signs of movement, and the speed gradually rises until an exceedingly rapid rate is reached. Similarly if pieces of gill are placed in a magnesium-free solution directly after excision, the *lateral* cilia do not stop but shew an increased rate of movement. By removing the magnesium therefore, the speed of the lateral cilia is increased, and eventually a rapid flickering movement sets in owing to a reduction in the amplitude. On replacing the magnesium, the rate quickly begins to fall, the amplitude increases, and the normal vigorous rhythm is resumed. It is clear that the presence of magnesium exerts a double effect on these *lateral* cilia, (*a*) it gradually brings them to rest, although the process takes some time, (*b*) it controls the rate at which they beat, by limiting their speed. Neither of these properties is shared by calcium, and the only metallic ion which can completely inhibit magnesium is that of potassium (Gray, 1926, 1), although it is probable that other metallic ions differ from magnesium only in degree.

The reaction of the *lateral* cilia to magnesium shews that the ions must be applied to the inner surface of the cell. Under normal conditions there is far too much magnesium in the sea-water to allow the cilia to be active, and yet in life the cells are active. It is only when the tissue is excised and the internal surface of the cells is bathed in sea-water instead of by blood that the magnesium affects the cilia. The blood contains more potassium and far less magnesium than sea-water, and it is the former fluid which controls the cilia. The point is important, for it indicates that metallic ions, like hydrogen-ions, operate inside the cell and not at its surface.

Summary of the effects produced by disturbances of ionic equilibria.

Any attempt to visualise the actual mechanism whereby such ions as hydrogen, potassium and magnesium exert their effects on the speed of the ciliary beat is rendered difficult by the complexity of the medium in which the changes occur. If we are right in assuming that cations can pass freely from the blood into the cells, then we may reasonably conclude that any change in the composition of the blood will induce in the cell such changes as occur in any other colloidal system. Unfortunately our knowledge of the behaviour of lyophil colloids to mixtures of electrolytes is very incomplete and does not carry us very far.

Being on the alkaline side of their isoelectric point, the intracellular proteins and other ampholytes will tend to react with cations; if exposed to a solution containing only one metal, *e.g.* sodium, the state of the system may be represented as follows:

$$\text{HROH} \rightleftharpoons \text{H}^{\cdot} + \text{ROH}' \qquad \ldots\ldots\text{(i)}.$$

$$\text{ROH}' + \text{Na}^{\cdot} \rightleftharpoons \text{NaROH} \qquad \ldots\ldots\text{(ii)}.$$

The lower the pH (*i.e.* the more alkaline the solution) the more completely will the ampholyte be ionised as ROH'; the more acid the solution the less will the ampholyte be ionised. At a certain pH all the colloid will be ionised, and if we then add an excess of sodium-ions by means of sodium chloride the amount of ionised ampholyte will be reduced and an unionised sodium salt will be formed. Most of the sodium salts of the proteins and other lyophilic colloids in the cell are dispersed in water or in a solution of sodium chloride and it is for this reason that the cells become unstable in such solutions or in those of other monovalent salts. The divalent salts have, however, very different properties. Magnesium salts of the proteins are ionised to only a limited extent, although they may or may not be freely soluble in water. If magnesium is therefore added to system (ii) there will be an increase in the amount of unionised protein,

$$2\text{ROH}' + \text{Mg}^{\cdot\cdot} \rightleftharpoons \text{Mg(ROH)}_2.$$

The effect of calcium is different, since at a pH comparable to that in the cell the calcium salts of the proteins are only soluble to a very limited extent,

$$2\text{ROH}' + \text{Ca}^{\cdot\cdot} \rightarrow \text{Ca(ROH)}_2 \text{ (precipitated)}.$$

The distinction between magnesium and calcium can be illustrated by the fact that the presence of magnesium in a system of alkaline albumen greatly inhibits the power of calcium to precipitate the colloid (Gray, 1926, 2). As already mentioned, the general effect of increasing the concentration of calcium within the cell is to convert any unionised protein salt into an insoluble calcium salt without at first affecting the concentration of free calcium-ions. The calcium compounds, although relatively insoluble, appear to be highly ionised.

The direct effect of ions on ciliary movements can be harmonised with their general effect on lyophilic colloids on the assumption that the ciliary mechanism is built up from (*a*) a solid protein phase rich in calcium and probably containing magnesium, (*b*) a liquid phase containing all the metallic ions, ionised protein elements, and unionised protein salts rich in magnesium. The effect of magnesium on the system is to keep the concentration of ionised protein in the liquid phase below that at which the calcium-ions effect precipitation, and if magnesium is removed the unionised protein molecules will dissociate until the precipitation level of Ca·· is reached. It has already been shewn that the presence of magnesium limits the speed at which the *lateral* cilia of *Mytilus* can move, and the suggestion arises that the limit of speed is determined by the degree of ionisation of some intracellular compound or surface. To some extent this hypothesis covers the facts. It accounts for the action of (a) hydrogen-ions, (b) magnesium-ions, (c) the difference between calcium and magnesium. At the same time it fails to give any adequate reason for the physiological differences between sodium and potassium and for the specific ability of potassium to antagonise magnesium (see p. 99).

Since the action of ionic disturbances on the stability of the intercellular proteins of the cell receives an adequate explanation by the hypothesis sketched above (Gray, 1926, 2), it seems reasonable to suppose that those changes which occur in the physical conditions of an extracellular protein will also occur in an intracellular protein if the conditions to which they are exposed are identical. For the moment therefore we may accept the view that the speed of the ciliary beat is influenced by cations by virtue of their power to influence the degree of ionisation of some essential element or elements in the cell.

References

Gray, J. (1922, 1). Proc. Roy. Soc. 93 B, *104*.
—— (1922, 2). Proc. Roy. Soc. 93 B, *122*.
—— (1924). Proc. Roy. Soc. 96 B, *95*.
—— (1926, 1). Proc. Roy. Soc. 99 B, *398*.
—— (1926, 2). Brit. Journ. Exp. Biol. 3, *167*.
Höber, R. (1914). *Physik. Chem. der Zelle und Gewebe, 508, 532*.
Kugelmass, I. N. and Shohl, A. T. (1924). Proc. Soc. Exp. Biol. and Med. 21, *6*.
Lillie, R. S. (1906). Amer. Journ. Physiol. 17, *189*.
Mayer, A. G. (1910). Carnegie Instit. Washington, Public. No. 132.

Chapter VI

THE METABOLISM OF CILIATED CELLS AND
THE NATURE OF THE CILIARY CYCLE

In the metazoa, ciliated cells appear to be restricted to aerobic conditions; in some protozoa, however, ciliary movement is believed to occur in the complete absence of oxygen (Pütter, 1904; Cole, 1921). That ciliary activity usually involves an associated consumption of oxygen by the cells is shewn by the fact that when the speed of the cilia is varied by appropriate means a corresponding variation occurs in the amount of oxygen consumed. A striking case is afforded by the lateral cilia on the gills of *Mytilus*. These cilia can readily be brought to rest in the excised gill (Gray, 1924), but if a small concentration of veratrin is added to the medium, the cilia rapidly acquire an active and prolonged period of movement. By measuring the oxygen consumption of the tissue in a suitable micro-respirometer the onset of movement is found to involve an increase of 45 per cent. in the oxygen consumption. The means employed to set up activity is immaterial; if the lateral cilia are activated by potassium-ions instead of by veratrin the oxygen consumption rises 47 per cent.; if a mixture of K· and hydroxyl-ions is used as an activating agent the oxygen rises 42 per cent. (Gray, p. 106). Similarly the cessation of ciliary movement caused by acid or by cold reduces the oxygen consumption.

Engelmann shewed in 1868 that a fragment of the ciliated epithelium of the frog, isolated in a drop of water, remained active for as much as two hours in an atmosphere of hydrogen. The conditions of his experiments did not shew, however, how far the tissue was drawing on the O_2 dissolved in the medium, and how far the cells were living anaerobically. The rate at which an undisturbed drop of water will part with its oxygen to an atmosphere of hydrogen is extremely slow, and consequently the survival time of cells isolated in such a way varies with the size of the drop. By using an indicator for free oxygen this factor can be eliminated and a striking experiment be performed as follows. To sea-water is added sufficient haemoglobin to allow a fairly thin film to give a clear oxyhaemoglobin spectrum when observed through a small Zeiss micro-spectroscope. A small fragment of *Mytilus* gill is then washed in the haemoglobin solution and placed in a gas chamber with as little free solution as possible. After turning on a current of pure but damp hydrogen, the time is recorded at which the haemoglobin is reduced, and the time at which the cilia cease to move, the latter observations being made through an ordinary eyepiece after removing the spectroscope. The following are the results of such an experiment:

TABLE XVI.

Time (mins.)	Atmo-sphere	Spectrum	Ciliary activity
0	Hydrogen	Oxy. haem.	Active
7	,,	,,	—
10	,,	(Indistinct)	—
13	,,	Reduced haem.	—
20	,,	,,	—
30	,,	,,	Slower than normal
50	,,	,,	Slow
60	,,	,,	Slow
61	Oxygen	Oxy. haem.	Active (recovery very rapid)
65	Hydrogen	,,	,,
185	,,	Reduced haem.	No movement
186	Oxygen	Oxy. haem.	,,
190	,,	,,	Partial recovery
250	,,	,,	Active but slower than normal

It is clear that ciliary activity occurs for some time after the concentration of free oxygen has been reduced to a low level. After prolonged absence of oxygen, recovery is slow and often not complete, and this is due to some change in the tissue and not to the time required for the penetration of O_2 into the depleted tissue (Gray, 1924). The conclusion that cilia can move for about half an hour in the absence of oxygen is confirmed by the fact that activity occurs for a similar period (30–45 mins.) in the presence of relatively high concentrations of cyanides (0·1 per cent. NaCN in sea-water) (Gray, 1924).

We can therefore conclude that ciliated cells like muscles and other tissues can perform their functions anaerobically but that oxygen is necessary for prolonged activity. At the same time it is very doubtful whether ciliated cells ever make any physiological use of their anaerobic powers. As far as we know, cilia cannot be

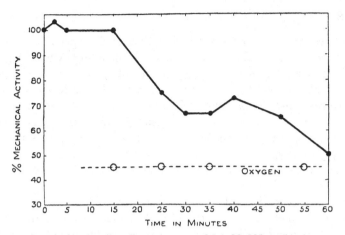

Fig. 72. Graph shewing the effect of 0·00005 Mol . NaCN on the oxygen consumption and on the mechanical activity of the frontal cilia of *Mytilus*. Note that the O_2 consumption falls very quickly after the application of cyanide, but the mechanical activity falls very much more slowly. (After Gray.)

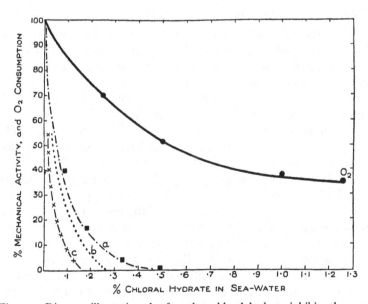

Fig. 73. Diagram illustrating the fact that chloral hydrate inhibits the mechanical movement of cilia much more readily than it affects the rate of oxygen consumption. *a*, Activity of frontal cilia. *b*, Lateral cilia. *c*, Latero-frontal cilia. (After Gray.)

fatigued and any process which quickens the speed must also increase the demand for oxygen to a corresponding extent. A prolonged lack of oxygen is rapidly detrimental, whereas a prolonged accumulation of CO_2 is readily reversible; the presence of CO_2, however, rapidly cuts down both activity and oxygen consumption, and this is the mechanism used by such animals as *Mytilus* when normal respiration is inhibited. By retaining CO_2 they conserve the supply of oxygen.

The facts already given shew that although the processes of movement and the processes governing oxygen consumption are undoubtedly correlated, yet there are two separate mechanisms either of which can be interfered with without affecting the other to a corresponding degree. Just as it is possible to cut down the O_2 consumption and leave the cilia active, so the activity can be cut down and still leave a considerable absorption of oxygen. This can be done by using narcotics such as chloral hydrate (see Fig. 73).

Analysis of the whole ciliary cycle.

Since the cilia on the gills of *Mytilus* are capable of expending energy for prolonged periods after excision from the animal, it follows that the substance from which this energy is derived must be present in the cells in relatively large amounts. Quite recently Mr E. Boyland has analysed the gills of *Pecten* and his preliminary results suggest that ciliary activity may involve the conversion of glycogen into lactic acid. In a sample of freshly excised gills 0·17 per cent. of the dry weight consisted of glycogen, whilst 0·08 per cent. was lactic acid; in corresponding gills · analysed 36 hours after excision the percentage of glycogen had fallen to 0·035, whereas that of lactic acid had risen to 0·13 per cent. These values are admittedly small, and are of the order which might be found in almost any type of tissue; at the same time, it should be remembered that a substantial fraction of the dry material in the gill represents, not ciliated cells, but the skeletal tubes. The results of further analyses will be awaited with interest, but it seems not unreasonable to conclude that cilia, like muscles, can derive their energy from glycogen.

It has been suggested that a glycoprotein forms the source of ciliary energy in *Mytilus* (Gray, 1924); this conclusion was based on three pieces of evidence: (i) the failure to detect significant amounts of carbo-hydrates in the cells, (ii) the presence of mucoid granules near the bases of the cilia, (iii) the respiratory coefficient. It would be of great interest to know whether carbohydrates can be detected in the gills of *Mytilus* if looked for by more refined methods.

The cilia on excised gills must cease to beat when the whole of the substance supplying their energy (G) has been used up. Under many conditions, however, the duration of beat does not depend upon this factor (Gray, 1926). The *lateral* cilia of *Mytilus* may come to rest in a very short time after their blood supply has been replaced by sea-water. If they are then exposed to veratrin, an excess of potassium-ions, or to a reduced concentration of magnesium-ions, activity is regained and, on transfer to normal

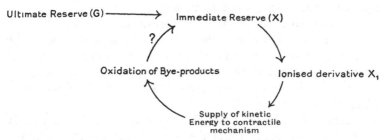

sea-water, is maintained for many hours. Hartree and Hill (1922) have shewn that veratrin acts on a muscle by increasing the amount of energy available for release when the muscle is stimu-lated. We may therefore assume that the drug exerts a similar effect on cilia and accelerates the conversion of glycogen into a substance (X) which can supply energy direct to the ciliary mechanism. When an excised gill is placed in normal sea-water the formation of X from G will cease, just as glycogen ceases to form hexosephosphoric acid when magnesium is present in a muscle (Embden, 1923). Under these conditions the cilia will continue to beat until no more intermediate substance (X) is present; this period may be very long as in frontal ciliated cells, or may vary with the degree of starvation of the animal as in the lateral cells (Gray, 1926).

The actual speed of the beat *rapidly* responds however to alterations in the nature of the cations (particularly hydrogen) present in the blood, and in the case of the *lateral* cilia to magnesium. These changes differ from those already mentioned in that they are extremely rapid, and as suggested elsewhere (p. 103) they appear to operate by changing the degree of ionisation of some essential element. Consider the case of a fragment of tissue in which a supply of X has been mobilised, sufficient to keep the lateral cilia moving for many hours at a rapid rate[1]. If magnesium is now added, the speed *at once* drops; or if hydrogen-ions are added the cilia at once stop. The speed of the beat is therefore susceptible to two types of inhibitory change, (*a*) one which takes place gradually and which may not be observable for some hours after the environment has been changed, (*b*) one which occurs at once. It is this latter change which is possibly to be associated with changes in ionisation (p. 103).

The third phase of the cycle involves the movement of the cilia themselves or what may be termed the contractile mechanism. The only known way of preventing this mechanism from using available energy is to deprive the cell of water (Gray, 1922). The nature of the machine is discussed elsewhere and is obviously obscure (see Chapter II).

Finally we must deal with the oxidative cycle. It has already been seen that this phase of the cycle probably occurs after the mechanical changes have been completed and not before. By assuming, for the time being, that the oxidative process tends to rebuild some of the immediate reserve which has been broken down, a reasonable explanation is provided for the gradual reduction in the speed of the beat during lack of oxygen or in the presence of KCN, and for the slowness of the recovery in speed when O_2 is again being absorbed. In this way the whole cycle of ciliary movement can be harmonised with that of a muscle cell. The maintenance of full oxidative activity in the absence of $Ca^{..}$ is not easy to understand. The slow reduction in mechanical speed would suggest that the concentration of X' was falling, but if this were the case, one would expect the oxygen consumption to fall. The full oxidative level, on the other hand, suggests that the energy derived from X' is being short circuited away from the contractile mechanism.

---

[1] By removing magnesium from the medium.

The same phenomenon occurs in the heart, and Mines (1913) suggested that calcium forms some essential part of the contractile machine. The objection to this view in the case of cilia is that there is no change in the form or amplitude of the beat; activity is simply reduced in speed till movement eventually ceases.

In the diagram below an attempt is made to indicate the points in the cycle at which various of the more powerful factors exert their influence.

Based on very limited data, the above analysis should not be regarded as anything more than a working hypothesis; at the same time it receives fairly strong support from a detailed comparison with other types of contractile tissue. This comparison

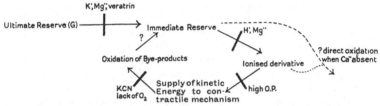

(The bars which cross the arrows indicate the point in the ciliary cycle at which the various reagents exert their characteristic effects.)

is discussed below, but one or two typical examples may be alluded to. Hartree and Hill (1922) have shewn that the drug veratrin greatly increases the amount of energy that is mobilised in striated muscle and which is ready for liberation when the muscle is stimulated. The application of the drug to the *lateral* cilia of *Mytilus* liberates enough energy to keep the cilia beating for many hours. The comparison of the two tissues obviously supports the view that the first step in the ciliary cycle consists in the conversion of some inactive chemical reserve into some such form as can supply energy to the cilia, and be regarded as the fuel for the contractile mechanism. From totally different arguments Embden (1923) concludes that in muscle the conversion of the inactive reserve of glycogen into the active fuel of the muscle is inhibited by magnesium, and we have seen that the metal also plays this rôle in ciliated cells. Step by step the five main processes of the contractile cycle appear to be the same in muscle and in cilia, although we have no direct evidence in the latter concerning (i) the nature of the substance which is oxidised as a result of activity, (ii) how much, if any, of this oxidation leads to a resynthesis of an active element or fuel. The hypothesis, as set forth, postulates that cations are able to diffuse from the blood into the cells and this does not appear to be true for all types of tissue. On the other hand, so many facts fall into line on the hypothesis that ciliary and muscular movement are of the same fundamental nature that we may accept the idea as something more than a speculative analogy.

Comparison of cilia with muscle.

The most obvious physiological parallels to the spontaneous rhythm so characteristic of cilia are presented by cardiac muscle cells in vertebrates and by certain locomotory muscles in the invertebrates. Morphologically, the ciliated cells stand apart from all others in that the contractile elements are restricted to one end of the cell, whereas in muscle cells either the whole of the cell or at any rate a considerable fraction of it participates in the change in form which accompanies contraction and relaxation. In detailed structure only one feature is common to all types, viz. there are usually present certain basophil structures not present in other types of cell; the arrangement of these basophil elements is, however, so variable in different types of muscle as to render their relationship to the contractile process too obscure for analysis.

Physiological comparison, on the other hand, leads quite definitely to the conclusion that the factors involved in both muscular and ciliary movement are very closely related, if indeed they are not identical. In making any useful comparison between the two types of cell, care must be taken to ensure that the fundamental conditions are identical. If we are dealing with an excised strip of ciliated epithelium with its characteristic automatic activity we must compare this with an excised strip of muscular tissue which also is automatically active. Failure to observe this obvious precaution has led to considerable confusion (see Mayer, 1910).

Perhaps one of the most striking properties common to both cilia and cardiac muscle is that neither tissue can readily be fatigued. As long as both are properly ventilated they may continue to function at an exceedingly rapid rate for prolonged periods without any diminution of activity. This feature may be correlated in the case of cilia with two facts, (a) the amount of work done per second is comparatively small when compared with striated muscle, (b) they do not accumulate an oxygen debt even at high speeds. The oxygen consumption runs parallel to the rate of beat in both the heart (Evans, 1912) and in cilia (Gray, 1923).

The parallel which exists between cardiac muscle and cilia when the temperature is varied has already been mentioned, the effects being similar both qualitatively and quantitatively. The action of hydrogen-ions is correspondingly similar if adequate care is given to the conditions under which the effect is observed. We have seen that if cilia are brought to rest by a minimum dose of acids, the effect of the latter is to slow both phases of the beat, and later to stop the cilia in the relaxed position. The same effects are observed in the sinus of the heart (Dale and Thacker, 1914), but in this case it is necessary to note that in addition to a retardation of the speed of the contractile cycle there also occurs a diminution in the frequency of the beats. A similar effect of acid on muscular contraction can also be seen in skeletal muscle. The following figure (Fig. 74) from Fröhlich (1905) shews clearly that in the presence of CO_2 both the rate of contraction and the rate of relaxation are decreased. In all types of contractile tissue the amplitude of the beats in the presence of CO_2 remains at its full value.

With stronger acids the sinus and cilia both stop in a permanently contracted state. With alkalies the rate steadily quickens in both cases up to about pH 9, but above that value the speed falls and the most obvious feature in both cases is the marked retardation in the rate of relaxation.

In dealing with the effect of other ions on the heart we meet with a difficulty, owing to the fact that comparatively few experiments have been made on excised tissue which is really automatic in its activity. Such observations as are available harmonise closely with those on cilia although in some cases a certain amount of selective comparison is liable to occur. In the first place both cardiac tissue (Sakai, 1914) and cilia (Gray, 1922) are comparatively insensitive to anions if the normal concentrations of cations are maintained. In both cases potassium increases the speed of the beat and may lead to prolonged but not permanent tonic contraction. The effect of an increased concentration of potassium (1 c.c. of 0·2 per cent. KCl in 10 c.c. Ringer) on the excised sinus of the frog is markedly to increase the rate of the beat (Köhn and Pick, 1920), just as is the case with all cilia. Still higher concentrations of potassium send the excised sinus into temporary tonic contraction; this is also the effect on the *latero-frontal* cilia of *Mytilus*. In both cardiac muscle and cilia there appear to be considerable variations in the sensitivity of individual tissues to upward and downward changes in the concentration of potassium.

We have seen that in an absence of calcium cilia come to rest in the relaxed position, although there is no diminution in the rate of oxygen consumption. Similar phenomena occur in the heart (Locke, 1907; Mines, 1913). The regulative action of magnesium on the speed of the lateral

G   8

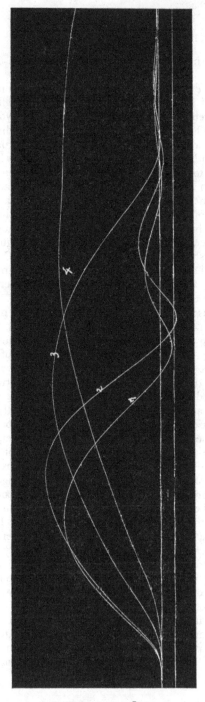

Height of Contraction

Time

Fig. 74. Effect of CO_2 upon the contractile cycle of an isotonic twitch of a frog's sartorius muscle. Note that the time occupied by contraction and by relaxation is gradually increased, but the amplitude of the contraction remains unchanged. (After Fröhlich.)

cilia of *Mytilus* is comparable to its action on the heart of *Pecten*. The indifference to magnesium of the *frontal* cilia is comparable to the state of affairs in the heart of the Octopus (Fredericq, 1913).

This list of parallel observations might be considerably extended, but not without danger. The reactions induced by ions in the different components of cardiac muscle are still far from clear (see Hogben, 1925); and the situation can only be reasonably satisfactory when we are in a position to compare the response of cilia to various conditions with the corresponding response of a rhythmical muscle from the same animal.

It is, however, possible to compare the reactions of cilia with the particular type of muscle found in certain Medusae. The reactions of the sub-umbrella tissue of *Cassiopea* were extensively studied by A. G. Mayer (1906), and give step by step a remarkable parallel to ciliated tissue. During normal life the medusa swims spasmodically by a series of contractions comparable to those exhibited by the cilia of Ctenophores or veliger larvae. If, however, the marginal sense organs are removed all rhythmical movements of the muscles cease in normal sea-water. They can, however, be initiated and maintained by altering the ionic equilibrium in the surrounding medium in precisely the same way as are the *lateral* cilia of *Mytilus*. "Discs without sense organs are actively stimulated into pulsation for a short time in all excess of potassium from sea-water + 0·25 per cent. K_2SO_4 to a pure solution of K_2SO_4 or KCl, isotonic with the NaCl of sea-water." The inhibitory effects of magnesium are also well seen; discs will pulsate rapidly in all solutions lacking magnesium at an abnormally high rate. Acids cause a decrease in the rate and energy of the beat, but do not change its form. High osmotic pressure, on the other hand, does not affect the rate although its form is irregular.

These and other points of similarity between muscle and cilia have been summarised elsewhere (Gray, 1924), and from them only two influences appear to be possible: (a) either there is some fundamental structure common to the two types of contractile mechanism, or (b) both muscular and ciliary movements are dependent on some common property of the cells concerned, although this property is not specific to contractile cells only.

References

Cole, A. E. (1921). Journ. Exp. Zool. 33, *293*.
Dale, D. and Thacker, C. R. A. (1914). Journ. of Physiol. 47, *493*.
Embden, G. (1923). Naturwiss. 2, *985*.
Engelmann, T. W. (1868). Jena Zeit. 4, *321*.
Evans, C. L. (1912). Journ. of Physiol. 45, *213*.
Fredericq, H. (1913). Bull. Acad. Roy. Belgique. *758*.
Fröhlich, F. W. (1905). Zeit. für Allg. Physiol. *288*.
Gray, J. (1922, 1). Proc. Roy. Soc. 93 B, *104*.
—— (1922, 2). Proc. Roy. Soc. 93 B, *122*.
—— (1923). Proc. Roy. Soc. 95 B, *6*.
—— (1924). Proc. Roy. Soc. 96 B, *95*.
—— (1926). Proc. Roy. Soc. 99 B, *398*.
Hartree, W. and Hill, A. V. (1922). Journ. of Physiol. 56, *294*.
Hogben, L. T. (1925). Quart. Journ. Exp. Physiol. 15, *263*.
Köhn, R. and Pick, E. P. (1920). Pflüger's Archiv, 185, *235*.
Locke, F. S. (1907). Journ. of Physiol. 36, *205*.
Mayer, A. G. (1906). Publ. Carn. Instit. Wash. No. 47.
—— (1910). Publ. Carn. Instit. Wash. No. 132.
Mines, G. R. (1913). Journ. of Physiol. 46, *188*.
Pütter, A. (1905). Zeit. f. Allg. Physiol. 5, *565*.
Sakai, T. (1914). Zeit. für Biol. 64, *1*.

Chapter VII

METACHRONAL RHYTHM AND CILIARY CONTROL

Metachronal rhythm.

A cursory examination of any active ciliated epithelium reveals the fact that although the cilia may be beating at the same rate they are not beating in the same phase. Any particular cilium is slightly in advance of the cilium behind it in the series and slightly

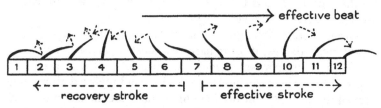

Fig. 75. Diagram to illustrate metachronal rhythm. Cilia 1 and 12 are at the end of the effective stroke; 2–7 indicate successive stages during the recovery stroke; 8–11 indicate stages during the effective stroke. All the cilia 1–12 are beating in sequence.

Fig. 76. Diagram illustrating the optical appearance given by a profile view of cilia beating in metachronal rhythm. (After Verworn.)

behind the one just in front of it. This regular sequence is known as *metachronism* and is observable in nearly all ciliated epithelia. Since all the cilia lying in the same line across the epithelium beat in approximately the same phase, regular waves of activity can be seen passing over the surface, thereby giving the well-known analogy to the waves which pass over a field of corn when exposed to the breeze. The crests of the metachronal waves are cilia at the zenith of their effective stroke; the troughs of the waves are cilia at or near the beginning of the recovery stroke.

The direction of the metachronal waves varies with different tissues. In the case of the frog's epithelium, or the frontal cilia of *Mytilus*, the waves move in the direction of the ciliary current, *i.e.* in the direction of the effective stroke of the cilia. In Ctenophores, however, the waves usually move in exactly the opposite direction. In these animals the effective stroke of the ciliated plates is in an aboral direction, so that the animals move mouth foremost. The metachronal waves, however, pass over the row

Fig. 77. Lateral epithelium of *Mytilus edulis* gill. The effective beat is in a plane at right angles to the plane of the paper: the metachronal wave passes in the plane of the paper in the direction of the arrow.

of cilia from the aboral to the oral end. In the *lateral* epithelium on the gills of *Mytilus* the metachronal wave moves at right angles to the effective stroke (Fig. 77).

Although the direction of the metachronal wave thus differs in different tissues, it is remarkably constant in each particular case. In the frog's oesophagus, von Brücke (1916) shewed that if portions of the epithelium were excised and replaced after being turned through 180° the original direction of the metachronal wave remained, although diametrically opposed to that of the rest of the epithelium. In *Mytilus* the lateral wave passes in opposite directions on the two sides of the gill-filament. It is only in certain Ctenophores that any evidence of wave reversal is found. In *Pleurobrachia* a rapid wave, on reaching the oral end of the ciliated row may be reflected aborally, but according to Parker (1905) it seldom passes over more than one-third of the whole row. Verworn (1890) induced reversed waves by stimulating the oral end of the row.

In the case of the Ctenophores (see below, p. 122) there is no doubt that, during life, the activity of the cilia is coordinated to some degree by definite 'sense' organs at the aboral end of the animal. At the same time metachronal rhythm is a fundamental property of ciliated epithelium, just as the power of automatic movement is a property of the individual cells. The beautiful rhythm of the *lateral* epithelium of *Mytilus* can be seen in gill fragments, whose proximal and distal ends have both been cut. The problem presented by the facts resolves itself into finding out how automatic units when side by side in a tissue beat in an orderly sequence.

An analogous problem is presented by certain flagellated cells, although the conditions are somewhat different. Numerous authors have observed that when the heads of individual spermatozoa are in intimate contact their tails beat synchronously, and a very striking example of the phenomenon can be observed in *Spirochaeta balbianii*. This organism is frequently found in the crystalline style of the oyster, and it moves by means of a clearly defined undulatory movement. In different individuals the number of waves passing down the body varies from three to about seven according to the speed of the animal. It frequently happens,

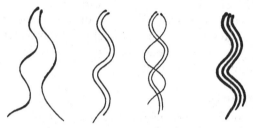

Fig. 78. *Spirochaeta balbianii* forming aggregates, the individuals in which soon establish synchronous movements.

however, that an organism attaches itself to the side of the style by one end, and continues to move the rest of its body in a lateral rhythm: if a second individual comes into the vicinity of a stationary individual it often attaches itself with its apical end in contact with that of its fellow. At first there is no coordination of body movements but within a few seconds complete coordination is established. In this way aggregates of several individuals are formed, all moving in perfect time. After a time the individuals of an aggregate can be seen to separate and resume independent movement.

The facts indicate that automatically contractile cells can coordinate their movements without the existence of any permanent or organic connection between the cells.

The behaviour of *Spirochaeta balbianii* finds an inorganic analogy in the evolution of hydrogen from chromium by means of hydrochloric acid. Ostwald (1900) found that hydrogen is evolved from isolated strips of the metal in regular pulsations, and that different pieces of metal exhibited different rhythms; when, however, the strips were in contact they all settled down into the same rhythm, only to lose it again when mechanical contact was broken. A mechanical model of temporary synchronised movement can be observed in the case of two pendulums. Two clocks, one of which is going slightly fast, and the other going slightly slow, when placed on the same stand will both keep correct time. Each clock transmits its vibrations to the stand and from the stand each clock receives vibrations set up by the other. The pendulum of the slow clock induces forced vibrations in the pendulum of the fast clock, causing the latter to increase its period of swing, and *vice versa*.

The problem presented by the metachronal rhythm of ciliated epithelia is more difficult, and it has so far received no adequate solution. The only line of attack appears to lie in an analysis of those extraneous factors which can modify both the speed of the beat and the transmission of the metachronal waves. It is obvious that the length of the metachronal wave must be proportional to the speed of the ciliary beat.

Excitation of cilia.

If a piece of frog's epithelium or gill of *Mytilus* be exposed to conditions (*e.g.* CO_2, KCN, cold) under which movement is extremely slow or just abolished, the cilia can be induced to beat more actively for a short time by mechanical irritation with a needle or other instrument. Kraft (1890) found that a mechanical stimulus thus applied to a localised region of a weak ciliary field induced activity along a band beginning at the spot stimulated and continuing for some distance in the direction of the normal metachronal wave (Fig. 79). The activity never extended far in the opposite direction or laterally from the point of stimulation. It is not clear from these experiments how far the main band of induced activity is the result of mechanical stimulus by the water current set up by the area (1234) subjected to direct stimulation[1].

Mechanical activation of stationary cilia also occurs in Ctenophores. If a complete ciliated comb is removed from *Pleurobrachia*, its cilia exhibit active movement for some time after excision, but after a time the movements gradually subside. From this condition active movement can again be evoked by mechanical agitation,

[1] See also Wyman (1925).

and a series of well-defined beats are observed in which the metachronal waves pass away from the direction of the effective beat.

It may be noted that these excitatory effects of stimuli are almost always restricted to tissues in a subnormal condition of activity and it may be doubted whether they ever occur in a normal epithelium *in situ*. Merton (1923) states that if a small fragment from the lips of the snail *Physa* be dissected out with the attached nerve the cilia soon come to rest unless the nerve is stimulated[1]. As Carter (1926) points out it is difficult to make sure that the

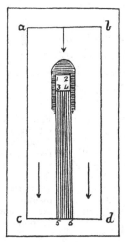

Fig. 79. Transmission of excitation over ciliated epithelium *abcd* of the frog's oesophagus. The area 1, 2, 3, 4 was stimulated mechanically. The shaded area shews the secondarily excited area. Note that the transmission is almost entirely restricted to the area 3, 4, 5, 6. The arrows indicate the direction of the effective beat of the cilia. (After Kraft.)

stimulus is conducted to the cilia by the nerve and not in a more direct manner. On the whole, the phenomenon of extraneous excitation of cilia recalls the effect of such treatment on moribund or quiescent cardiac muscle: in both cases the healthy tissue is automatically rhythmical, but when this property is lost for any reason it can be induced to reappear for a short time by providing an external stimulus.

Inhibition of cilia.

Whereas excitatory effects of external stimuli are comparatively rare, it is by no means uncommon to find clear instances of ciliary inhibition in perfectly normal tissues either *in situ* or after excision. Locomotory cilia are almost invariably under the control of the animal and this control is often, if not always, of an inhibitory nature.

[1] See also footnote on p. 4.

Jennings (1915) noted that when a *Paramecium* comes into contact with any rough surface (such as that of filter paper) the organism eventually comes to rest owing to the fact that the cilia in contact with the filter paper are no longer in motion; those over the rest of the body are greatly depressed and may also become motionless.

In view of the more recent observations of Saunders (1925) there is room for some doubt concerning the nature of the motionless structures seen by Jennings. Saunders has shewn that on coming into contact with a solid surface *Paramecium* frequently attaches itself by extension of its trichocysts; this process does not markedly interfere with the ciliary beat. At the same time, the description given by Jennings of the behaviour of ciliates in general, makes it almost certain that, under normal conditions, the activity of cilia can be reduced by some type of inhibitory control (see also p. 128).

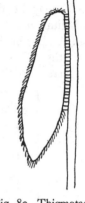

Fig. 80. Thigmotactic inhibition of cilia in *Paramecium*. Note that the cilia, in contact with the surface of the fibre, are at rest, whereas those on the remainder of the body are in motion. (From Jennings.)

The clearest cases of inhibition are those found in the higher invertebrata. The ciliated larvae of annelids, molluscs and polyzoa are all capable of bringing their cilia to rest, and they do so at irregular intervals or when externally stimulated. A typical instance has recently been investigated by Carter (1926) in the velar cilia of nudibranch larvae. During life these cilia shew well-marked alternations of rest and activity; if, however, a portion of the velum be excised from the animal, the cilia on this portion cease to shew any interruption of activity and beat steadily for many hours. Carter has shewn that narcotics used in such concentrations as are necessary to anaesthetise nerves abolish the periods of ciliary rest in the intact larva[1]. Taken in conjunction with the fact

---

[1] Similar results were obtained with other types of larvae by Merton (1923 *a*), but Carter was the first to demonstrate the presence of the nerve fibres. The power of ciliary inhibition is also lost by the Ctenophore *Beröe* when the animal is exposed to 0·2 per cent. chloral hydrate (Göthlin, 1920).

that nerve endings can be traced to intimate association with the ciliated cells, the evidence strongly suggests that the normal periods of rest shewn by the velar cilia are due to nervous inhibition.

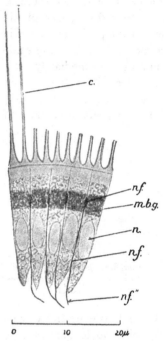

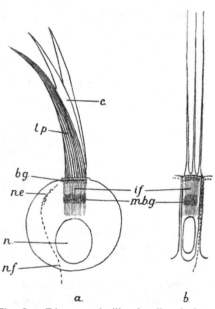

Fig. 81. *Archidoris tuberculata.* Living cells of the velar cilia stained *intra vitam* with methylene blue shewing nerve fibrils. *n.f.,* nerve fibrils between the ciliated cells; *m.b.g.,* granules; *n.,* nucleus; *n.f.″,* nerve fibrils passing into the tissues of the body; *c.,* cilium. (After Carter.)

Fig. 82. Diagram of ciliated cells of the molluscan velum. *a.* Side view of cilium. *b.* End view. Note the compound nature of the cilium. *t.p.,* triangular plates of cilium in side view; *b.g.,* basal granules; *i.f.,* internal fibres; *m.b.g.,* granules which stain with methylene blue; *n.,* nucleus; *n.f.,* nerve fibrils between the ciliated cells; *c.,* cilia. (After Carter.)

Another interesting example of 'controlled' locomotory cilia is found in the snail, *Alectrion trivitta* (Copeland, 1919). This animal moves by means of its ciliated foot; when the animal is at rest the cilia are motionless, but progression is entirely due to the activity of the cilia. Copeland found that when a resting animal is stimulated by touching one of its tentacles with a piece of fish meat the proboscis is extended and the pedal cilia begin to beat. At first sight this appears to be a reflex response involving the excitation of the cilia by some motor element. The phenomenon is not simple, however, for ciliary movement is greatly reduced or ceases altogether when the proboscis is worked over the surface of the foot as is often the case. On withdrawal of the proboscis the cilia shew increased activity[1]. On excising the foot, exactly the same series of events occur as in the Ctenophores. Immediately after excision, the cilia of *Alectrion* are motionless, but after a brief interval short outbursts of ciliary movement occur simultaneously with muscular twitchings. Gradually these periods of ciliary and

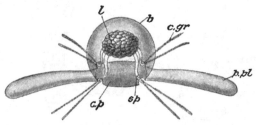

Fig. 83. 'Sense organ' of the Ctenophore *Hormiphora plumosa*. *b.*, bell; *c.p.*, ciliated plate; *c.gr.*, ciliated groove leading to locomotory ciliated plates; *l.*, lithites; *p.pl.*, polar plate; *sp.*, spring. (From Parker and Haswell.)

muscular activity become more and more frequent, and during the periods of muscular quiescence the cilia no longer cease to beat, although they beat more slowly than during muscular activity. Finally, after some hours, when all muscular movement has ceased, the cilia are found to be beating regularly at a constant rate which continues as long as the preparation remains intact.

In Ctenophores ciliary control is well defined. All these animals possess an aboral 'sense organ' (Fig. 83) from which finely ciliated grooves pass to the aboral end of each row of ciliated plates. In *Pleurobrachia* the normal mode of progression is with the mouth forward and with the contractile tentacles extended posteriorly.

[1] A similar phenomenon occurs in Ctenophores. When the tentacles of *Pleurobrachia* are extended, the ciliary movements are usually slow, when the tentacles are withdrawn, a marked increase in activity is sometimes seen in the ciliated combs.

If the animal is given a slight blow the ciliated plates instantly become motionless, the tentacles contract, and the organism sinks slowly in the water. After a short pause the cilia begin to beat again, the tentacles relax, and progressive movement is resumed. A marked case of mechanical inhibition has been described by R. S. Lillie (1906) in the genus *Eucharis*.

It should be remembered that none of the above tissues are exceptions to the rule that all ciliated cells are fundamentally capable of automatic activity. Excised portions of the cuticle of protozoa, portions of the velum or foot of molluscs, and the combs of Ctenophores all shew active and prolonged movement. Parker (1905) found that excised fragments of the combs of *Mnemiopsis* beat rhythmically for one or two days, and even isolated ciliated plates are active. Similar facts are recorded in other tissues by numerous observers. As far as is known normal ciliary inhibition is almost restricted to locomotory cilia and is rare in other types[1].

It seems clear that three distinct types of ciliary movement exist: (*a*) cilia which are normally in an active state of movement independent of any obvious external stimulus, (*b*) cilia which are motionless or only feebly active except when a stimulus is applied, (*c*) cilia which are active but can be brought to rest by some type of inhibitory control. The parallel to cardiac muscle is again striking. In both tissues one condition is fairly readily convertible into another except from type (*a*) to (*c*).

At present these facts do not throw any very clear light on the nature of metachronal rhythm. That an excitatory wave may pass over a region of motionless cilia (Kraft, 1890) indicates that during activity some form of disturbance passes down a line of cells along a different path from that occupied by the contractile elements. This independence of motivity and power to transmit a metachronal wave is not restricted to the frog; it is also seen in the Ctenophore *Mnemiopsis*, where transmission waves pass quite regularly over a series of plates which shew no indication of any movement (Parker, 1905). On the other hand Verworn (1890) found that if a series of plates are excised from *Beröe*, and one of the

[1] According to Fedele (1923) the branchial cilia of *Dolcolium* are under the control of the animal.

plates in the row be prevented from movement by some mechanical obstruction, the waves coming from the aboral end of the fragment are unable to pass the obstructed cilium, so that all the cilia lying on its oral side cease to move until the obstructed cilium is released. The difference between these reactions may be one of degree if there is any decrement in the waves as they pass along the conducting path. In the frog or in *Mnemiopsis* the wave on emerging from the quiescent region may be sufficient to control the orally situated cilia, whereas in *Beröe* it may not be sufficiently powerful. In all such cases it looks as though we were dealing with two distinct forms of control: (*a*) a localised inhibition which is not propagated, (*b*) the propagation of a general disturbance which is responsible for metachronal rhythm.

It seems fairly certain that a metachronal wave cannot pass over a severed strip of epithelium. In Ctenophores the initial effect of cutting one of the combs is to bring all the cilia to rest. It seems as though some general inhibitory mechanism were called into play; the same mechanism, in fact, whereby the cilia stop when the intact animal is touched. A short time after a comb is cut, however, both the aboral and oral parts begin to move (the aboral part usually begins first), but the two regions now beat independently.

Summary.

It is not easy to summarise the facts, and even more difficult to analyse the nature of ciliary control. As far as one can see the following types of ciliary disturbance exist.

(i) *Generalised inhibition*, which involves the use of some nervous or neuroid mechanism. Thus in the molluscan veliger the ciliary nerves convey an inhibitory stimulus direct from the central nervous system to the contractile elements. In Ctenophores we have to postulate a neuroid transmission over the whole surface of the animal which reaches the cilia via the aboral organ. In spite of much investigation, no trace of ciliary nerves has been found in these animals. The part played by the aboral organ appears fairly definite. It communicates with the locomotory cilia by finely ciliated grooves (see Fig. 83), and if the aboral sense organ of *Pleurobrachia* is destroyed by puncture with a needle two significant changes occur, (*a*) the power of general inhibition is lost [1], (*b*) the eight ciliated combs are no longer coordinated. In this way the animal loses its power of directional movement, and usually rotates rapidly on its own axis. It seems clear therefore that generalised inhibition is something impressed upon the cilia along definite tracts and by the intervention of the aboral organ.

(ii) *Localised inhibition*, which is the result of a stimulus applied at any point of the ciliated epithelium. As a rule localised inhibition is restricted to the point of application of the stimulus, although in *Beröe* it may spread in an oral direction. Whether localised inhibition is effected by the same mechanism as generalised inhibition is difficult to determine. In *Beröe* the facts would be explicable on the assumption that, along the efferent tract from the apical organ, a disturbance can only be propagated in an oral direction, so that both generalised and localised inhibition might be due to the same system of control. But in other Ctenophores we have seen that whereas a generalised inhibition spreads very rapidly along the efferent tract, a localised inhibition does not spread at all.

It is probable that both generalised and localised inhibitions are only effective if the automatic power of movement of the cells is below a certain level. Thus if the speed of the velar cilia of veliger larvae is increased by the presence of potassium or by veratrin, normal generalised inhibition does not occur (Carter, 1926); in *Eucharis* the automatic speed can be increased by reducing the concentration of calcium (Lillie, 1906) and mechanical localised inhibition no longer takes place.

(iii) *Excitatory control*. This, we have seen, is restricted to cilia whose automatic powers are extremely feeble. The only natural case is that described by Merton on the lips of *Physa*, although it is a general property of all moribund ciliated tissues.

(iv) *Metachronal control*. The facts described all point to the conclusion that the power of automatic movement is a fundamental property of all healthy ciliated cells, and that the control, if any, to which they are subjected by nervous or other external influences

[1] The experiments of Fedele (1924) shew that this is not the case in certain Neapolitan genera of Ctenophores; in these forms the normal response to mechanical stimulation appears to be effected by neuroid changes which are independent of the sense-organ.

is of an inhibitory nature. We have therefore to consider the mechanism whereby the cilia of an excised strip of tissue do not beat independently but in definite correlation with each other.

Most of the suggestions hitherto put forward have been based on the assumption that the typical cilium is not automatically active and that metachronal movement is due to the passage of a series of excitatory stimuli moving from one end of the series to the other. This appears to be the view of Parker (1905). The conception meets with two difficulties: (a) what is the source of these stimuli? (b) why are isolated ciliated cells always in active motion? Parker's view implies that isolated cells are active because they are continuously stimulated: we can, with equal justice, say that stationary cilia do not move because they are inhibited. We can only choose between these views by reference to the facts. Since velar cilia are continuously active in the presence of anaesthetics it is clearly of advantage to adopt the latter view.

The regulation which gives rise to the metachronal rhythm may quite likely be (and even probably is) independent of mechanical activity, but nevertheless it may be associated with other physiological activities of the actively moving cells. In a preparation of *Mytilus* gill it can often be noticed that the metachronal rhythm becomes irregular when the speed of the ciliary beat is low. If the *lateral* cilia have come to rest and the process of activation by veratrin be observed, it may be seen that the metachronal rhythm becomes more and more perfect as the speed increases. Although the mechanical activity may be locally abolished in such forms as *Mnemiopsis* we do not know that other expressions of cell movement of an electrical or other nature are not still in full activity, although the mechanical response is absent. All that we can suggest is that when a ciliated epithelium reaches a critical degree of activity there is set up a timing mechanism which travels by some other path than the visibly moving elements; concerning its essential nature we are entirely in the dark. It is conceivable that the timing mechanism consists of a series of inhibitory impulses passing along the tissue and that an equilibrium is set up between the automatic rhythm of the individual cells and the inhibitory impulses. If every cell is at the extreme end of its active cycle when the height of the inhibitory impulse reaches it, then normal metachronal rhythm will ensue; if a particular cell is not at rest at this moment the impulses will compel it to reach this position and so maintain the steady rhythm. Quite recently Grave and Schmidt (1925) have claimed that a series of intercellular fibrils can be detected in the epithelia on the gills of molluscs and have suggested that these represent an essential part of the coordinating mechanism. Bhatia (1926) has re-investigated these tissues, however, and his observations cast considerable doubt on the existence of any organised intercellular fibrils.

Verworn suggested many years ago that each cilium was mechanically stimulated by its active neighbour. This can hardly be the case since, as Parker pointed out, in many forms the effective beat and the metachronal wave travel in opposite directions.

Whatever be the mechanism whereby metachronal rhythm is effected, its physiological significance is fairly clear. A series of cilia all beating in the same direction, but independently of each other, could only produce a spasmodic flow of water; by beating metachronally a steady flow is ensured just as in a centrifugal pump or the paddle box of a steamer.

References

Alverdnes, F. (1922). Arch. Entw. Mech. 52, *281*.
Bhatia, D. (1926). Quart. Journ. Micro. Sci. 70, *681*.
von Brücke, E. T. (1916). Pflüger's Archiv, 48, *149*.
Carter, G. S. (1926). Brit. Journ. Exp. Biol. 4, *1*.
Copeland, M. (1919). Biol. Bull. 37, *126*.
—— (1922). Biol. Bull. 42, *132*.
Fedele, M. (1923). Public. della Staz. Zool. di Napoli. 4, *129*.
—— (1924). Public. della Staz. Zool. di Napoli. 5, *275*.
Göthlin, G. H. (1920). Journ. Exp. Zool. 31, *403*.
Grave, C. and Schmidt, F. O. (1925). Journ. of Morph. 40, *479*.
Jennings, H. S. (1915). *Behaviour of the Lower Organisms*. New York.
Kraft, H. (1890). Pflüger's Archiv, 47, *196*.
Lillie, R. S. (1906). Amer. Journ. Physiol. 16, *117*.
Merton, H. (1923). Pflüger's Archiv, 198, *1*.
—— (1923 a). Biol. Zentralb. 43, *157*.
Ostwald, W. (1900). Zeit. f. physik. Chem. 35, *33*.
Parker, G. H. (1905). Journ. Exp. Zool. 2, *407*.
Saunders, J. T. (1925). Proc. Camb. Phil. Soc. Biol. Series, 1, *249*.
Verworn, M. (1890). Pflüger's Archiv, 48, *149*.
Wyman, J. (1925). Journ. Gen. Physiol. 7, *545*.

Chapter VIII

THE DISTRIBUTION OF CILIA AND THE FUNCTIONS WHICH THEY PERFORM

Vibratile organs are found in all groups of the animal kingdom with the exception of the Nematoda and Arthropoda[1]. Not unnaturally the part played by ciliated surfaces is most obvious in aquatic animals whose movements and other activities are essentially slow. In rapidly moving vertebrates, whose tissues are characterised by a high rate of metabolism, the transport of food and other substances is carried out by powerful muscular elements, and cilia are restricted to a few comparatively insignificant functions. In many invertebrates, however, the mode of life is associated with a much lower rate of metabolism, and we find that cilia are at least as widely distributed and at least as significant as muscle fibres.

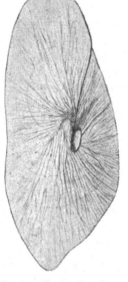

Fig. 84. *Paramecium* shewing neuromotor centre lying near the cytopharynx. Note the fibrils which radiate outwards. (After Rees.)

Protozoa.

In this group is found a wide variation in the form of the vibratile organs, and many examples have already been mentioned. Within recent years considerable attention has been paid to the mechanism which controls the movement of some of the commoner Ciliates. For example, in *Paramecium* (Rees, 1922), the neuromotor mechanism consists, in addition to the cilia, of two peripheral whorls of fibres (the larger on the oral side), and of two other groups of fibres associated with the cilia on the cytopharynx.

---

[1] At one time cilia were believed to exist in the larva of *Chironomus*, but this has never been confirmed. They undoubtedly occur in *Peripatus*.

All these four sets of fibres ramify out from a 'neuromotor centre,' lying near the anterior end of the cytostome. If the animal is

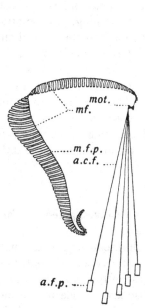

Fig. 85. Neuromotor apparatus of *Euplotes*. *mf.*, membranella; *m.f.p.*, membranella plates; *mot.*, motorium; *a.c.f.*, anal cirrus fibre; *a.f.p.*, anal cirrus plates. (From Taylor.)

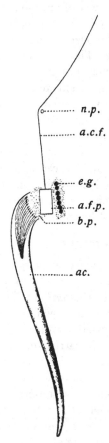

Fig. 86. Anal cirrus of *Euplotes*. *a.c.f.*, anal cirrus fibre ending in basal plate (*a.f.p.*); *e.g.*, ectoplasmic granules; *b.p.*, basal granules; *ac.*, cirrus. (After Taylor.)

injured in the neighbourhood of the neuromotor centre a distinct interruption of coordinated movement results. If the cytopharyngeal fibres are cut a loss of coordination occurs between the cilia behind the cut, and those in front of it. It seems reasonable to regard these fibres as conductile in function, and as capable of controlling the movements of the cilia, although the latter retain their power of autonomous movement.

An interesting experimental analysis of neuromotor mechanism and of movement has been carried out by Taylor (1920) in *Euplotes patella*. This organism usually creeps over the surface of the container, but it is also capable of swimming. Locomotion by creeping is effected by means of the cirri on the ventral surface aided by the membranellae. If the internal fibrillar system which radiates out from the neuromotor centre (Fig. 85) is cut, co-ordination between the movements of the ventral cirri and those of the membranellae breaks down. The effective stroke of the membranellae can apparently be reversed, as is also the case of the anal cirri. By an orderly and coordinated variation in the movements of the different cirri and membranellae, nine different types of locomotion can be carried out.

Coelenterata.

The ciliary surfaces of Actinians were investigated by Carlgren (1905) who described four main types.

(i) In *Protanthea* (Fig. 87) the adult resembles the larva in being ciliated over its whole external surface: the cilia on the outer surface beat upwards, and on reaching the base of the tentacles some of the current is carried on to the oral disc and thence to the mouth. The rest of the current passes along the tentacles to their tip. When the animal is feeding, the whole of the ciliary currents are directed to the mouth since the tentacles are bent over as shewn on the left of Fig. 87. When the tentacles are extended any foreign particle falling on their surface is swept to the tips and so removed from the animal. As far as is known there is no organised exhalent ciliary current leaving the cavity of the animal and it is not known how indigestible particles are removed.

(ii) In *Metridium* or *Sagartia* only the disc, tentacles, and stomodaeum are ciliated. When the animals are feeding a strong inward current exists at the mouth (Fig. 88) which sweeps particles inwards. When 'at rest' the oesophageal current, except on the siphonoglyph, is directed outwards and particles are thereby removed from the surface of the animal.

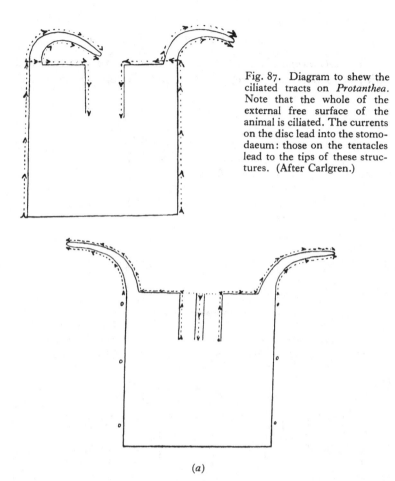

Fig. 87. Diagram to shew the ciliated tracts on *Protanthea*. Note that the whole of the external free surface of the animal is ciliated. The currents on the disc lead into the stomodaeum: those on the tentacles lead to the tips of these structures. (After Carlgren.)

(a)

Fig. 88 (a) and (b). Diagrams of ciliary currents in *Metridium*. Note the restriction of cilia to the tentacles, oval disc, and stomodaeum. Note that the figures indicate a reversal of the stomodaeal current during feeding. The siphonoglyph current is always directed inwards. In position (a) the animal is not feeding. In position (b) food is being carried into the gastric cavity. (After Carlgren.)

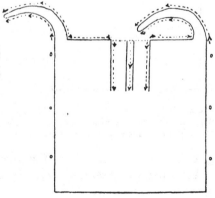

(b)

(iii) In *Caryophyllia* and *Taelia* the tentacles are heavily armed with nematocysts, and are, therefore, not unnaturally, unciliated.

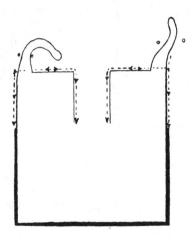

Fig. 89. *Caryophyllia*. Note that the tentacles are non-ciliated. They are richly armed with nematocysts. (After Carlgren.)

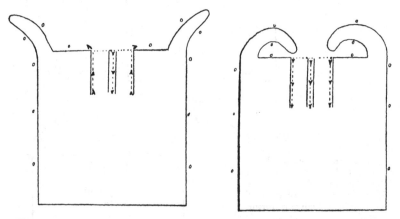

Fig. 90. *Taelia*. Note the restriction of cilia to the siphonoglyph and stomodaeal walls. The tentacles are heavily armed with nematocysts. Cilia are only of very secondary importance in feeding. (After Carlgren.)

The ciliary currents, as seen by Carlgren are shewn in Figs. 89 and 90. According to the same observer the direction of the stomodaeal current is reversible in *Taelia*. In both *Caryophyllia* and *Taelia*, the functions of the cilia are apparently reduced to the minor rôle of dealing with food particles brought to the mouth by the muscular movements of the tentacles.

The ciliation of *Metridium marginatum* has been described by Parker (1896) and is of peculiar interest. If a small amount of powdered carmine in sea-water be dropped on to the tentacles of an expanded anemone, a momentary waving of the tentacles is observed. This is followed by a period of quiescence, during which the carmine, matted together by mucus, is slowly carried from the bases of the tentacles towards their tips. If the tentacles concerned are on the outer edge of the tentacular zone, the carmine on reaching the end of the tentacles falls off beyond the oral disc. If, on the other hand, the tentacles in question do not reach the edge of the disc, the carmine falls on to other tentacles which transport it further towards the outer edge, till eventually the disc is freed from the foreign particles altogether. The movement of the carmine on the tentacles is due to cilia which are constantly in motion and which are sufficiently strong to propel an excised tentacle over the surface of a dish. If instead of carmine a small fragment of crab's muscle be dropped on to the tentacles, the tentacles near the fragment contract violently on the side next to the fragment so that the latter is moved nearer to the mouth, and by the tentacular cilia it is eventually deposited on the lips of the stomodaeum. The stomodaeum has two sets of cilia: (i) those in the siphonoglyph which invariably sweep particles into the gastro-vascular cavity, (ii) the cilia on the lips which beat inwards when stimulated by food or KCl, and outwards at other times (see p. 57). In this connection, however, the observations of Elmhirst (1925) on *Actinoloba* are of importance. "If small examples of *Actinoloba dianthus* are watched under the microscope, the use of carmine will shew that a steady centrifugal ciliary flow passes along the bases of and up the tentacles. For example, a drop of carmine and water placed over the mouth will gradually disperse centrifugally and fall off the tentacle tips. On the other hand, a suspension of food in water, *e.g.* diatom culture or carmine and crushed mussel, will be seen to be drawn down into the mouth— any particles falling within reach of the outgoing tentacular current being swept away. When the anemone is expanded the effect of the centrifugal flow along the numerous tentacles is to cause a compensating current to set in towards the middle of the disc, *i.e.* the mouth, and any suitable particles so brought in are usually swallowed (p. 151)...Longitudinal grooves run down the gullet and when food is being swallowed the inflow is along the grooves;

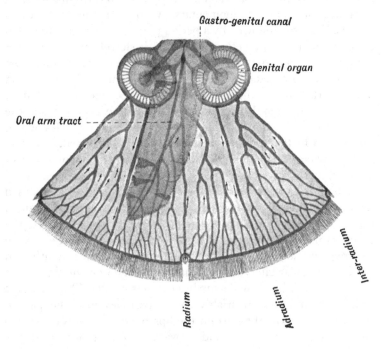

Fig. 91. Diagram shewing the course of ciliary circulation in the radial canals of *Aurelia*. (After Widmark.)

conversely a ciliary outflow runs up the ridges, for example, when a bolus of waste is discharged it is passed out by the cilia on the ridges aided by a certain amount of contraction of the stomodaeal wall. At times there is a vortex in the gullet when both sets of cilia are in action at once" (p. 152).

In Actinians there can be little doubt that the surface ciliary currents are of respiratory significance and are instrumental in maintaining the surface against ectoparasitic infection. In some forms there is a definite nutritive value in some of the currents,

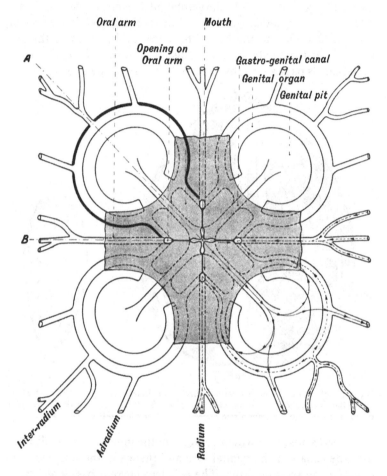

Fig. 92. Diagram shewing the course of ciliary circulation in the genital pits and other organs of an adult *Aurelia*. (After Widmark.)

which in others has been partially or completely replaced by muscular mechanisms.

The common Schyphomedusan, *Aurelia aurita* (Figs. 91 and 92), provides an interesting series of surface and tubular currents. The internal currents have been investigated by Widmark (1913). The inhalent current enters the animal by the mouth and passes up the four gastrogenital canals to the genital pits. Some of the water then passes up the adradial canals, and on reaching the circular canal, diverges to each side and begins to return to the centre of the disc by the radial and inter-radial canals. The water which enters the radial canals passes straight back to the exhalent opening at the base

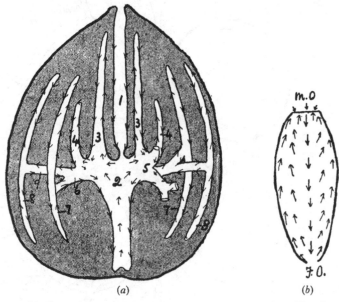

(a)   (b)

Fig. 93. (a) General course of ciliary circulation in the Ctenophore *Pleurobrachia*. (b) Ciliated tracts on the stomodaeum of *Pleurobrachia*. m.o, mouth; F.O. opening of funnel. Note antagonistic currents. (After Gemmill.)

of the oral arms; that which passes into the inter-radial canals flows into the canals of the genital sacs and thence to the four exhalent openings on the oral arms. The exhalent current passes along the surface of the oral arms and is therefore discharged well away from the inhalent current entering the mouth. That this circulation is maintained by cilia is shewn by the fact that it is exhibited by animals in which all muscular movement has been abolished by narcotisation with ether. The course of the circulation can be watched by means of Indian ink particles, and one complete circulation occurs every twenty minutes. The ciliary currents on the surface of an adult *Aurelia* have been described by Orton (1922), and on the larval ephyra by Gemmill (1921). In general the currents on the umbrella surface are centrifugal, whilst on the sub-umbrella surface they appear to differ in the two cases.

The locomotory cilia of Ctenophores have already been described, but there are also well-defined ciliated tracts in the stomodaeum and internal cavities (Gemmill, 1915). The course followed by small particles (see Fig. 93) appears to shew that each of the internal canals has an ingoing current on one wall and an outgoing current on the other. On the walls of the peristomium are two parallel but opposite currents (see Fig. 93 *b*). The general course of the circulation is interesting, for if Gemmill's observations are confirmed, it would appear that both inhalent and exhalent currents are organised along definite ciliary tracts throughout their whole length; and that this may be associated with the fact that circulation must be maintained in narrow tubes closed at one end.

Echinoderms.

Among the Echinoderms ciliated larvae are extremely common, and Gemmill (1918) has shewn that nearly every organ and every surface in the body of an adult starfish is covered with cilia. There can be little doubt that in addition to removing debris from the surface, the external ciliation provides for the respiration of the superficial nerve tracts, since these areas are far removed from the perihaemal spaces. The currents along the ambulacral grooves being centripetal, fresh oxygenated water is constantly reaching the nerve ring, although the disc may be wholly or partially buried in sand (*Astropecten*). In *Porania* and *Solaster* the skin on the aboral surface is ciliated in such a way as to drive particles of debris in the direction of the anus, and there throw them up in a perpendicular stream away from the animal.

The ciliation of the perivisceral cavity produces a constant and complete mixing of the coelomic fluid in the interior of the disc and the anus, and by the same means circulation is maintained inside the gills.

In all starfishes ciliary currents pass from the mouth to the anus, but by various modifications there occurs a mixing of the fluids within the main gastric cavities, and in addition there is a radial circulation in the caecal outgrowths. Most Asteroids do

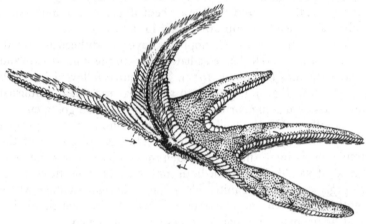

Fig. 94. *Astropecten* shewing the ciliary currents over the surface. (After Gemmill.)

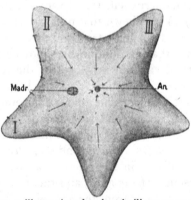

Fig. 95. Diagram illustrating the aboral ciliary currents in *Porania*. *An.*, anus; *Madr.*, madreporite. (From Gemmill.)

not depend on ciliary currents for obtaining their food, but in *Porania* this mode of feeding occurs (see Fig. 95). For an extensive account of the ciliation of other groups of Echinoderms, the monograph by Gislén (1924) may be consulted.

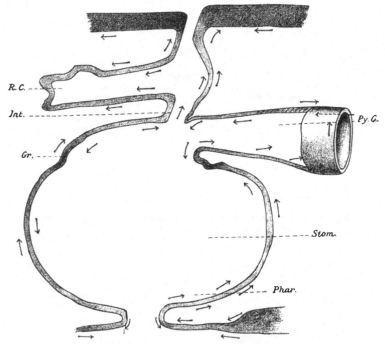

Fig. 96. Vertical section through stomach of *Porania* shewing ciliary circulation in the gut and splanchnopleural surfaces. *Py.C.*, pyloric caecum; *R.C.*, rectal caecum; *int.*, intestine; *Stom.*, stomach; *Phar.*, pharyngeal portion of gut; *Gr.*, groove between stomach and pyloric sac. (After Gemmill.)

Mollusca.

The course of ciliary circulation in the Mollusca has been followed in greater detail, perhaps, than that of any other group (Wallengren, 1905; Orton, 1912; Kellogg, 1915; Yonge, 1926), and since *Mytilus* has provided the material for some of the experimental work described in this book, it is convenient to take this animal as an example of Molluscan ciliation in general. An additional reason lies in the fact that the currents can readily be demonstrated to a class of students.

If a cluster of mussels are placed in clean sea-water it will be noticed that in each animal the shells soon begin to open and the edge of the mantle is slightly protruded; on putting a suspension of fine carmine particles into the water near the rounded end of

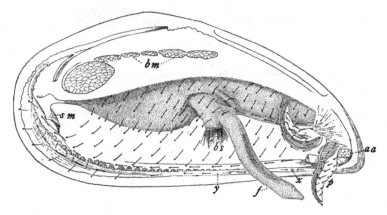

Fig. 97. *Mytilus edulis* shewing the exhalent currents on the mantle:
also currents on the labial palps. (After Kellogg.)

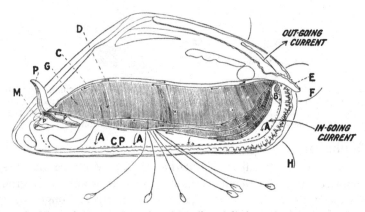

Fig. 98. View of the mantle cavity of *Mytilus edulis* from the left side to shew
the food-currents. *CP*, the dotted arrows and line at the ventral edge of the
mantle indicate the ciliated path which carries the material rejected by the
palps and that collected from the mantle to the point indicated by the arrow
above *B* in the figure. Here the rejected material is pushed into the exhalent
current. *A*, arrows indicating the paths of the heavier particles settling out of
the main food-stream. *B*, projection from the dorsal wall of the inhalent
aperture. *C*, line of attachment of the mantle to the body wall. *D*, arrows in
the supra-branchial chamber indicating the direction of the exhalent current.
F–H, points between which the main food current is drawn into the mantle
cavity. *G*, left outer gill lamella. *P*, *P'*, labial palps. *M*, position of mouth.
(After Orton.)

the animal two well-defined currents can be observed. The lower of these currents (Fig. 98) is directed inside the shells into the mantle cavity; the upper current leaves the mantle cavity through an aperture, known as the exhalent siphon, which is formed by the edges of the two sides of the mantle. The course of circulation within the mantle cavity is best seen by cutting through the

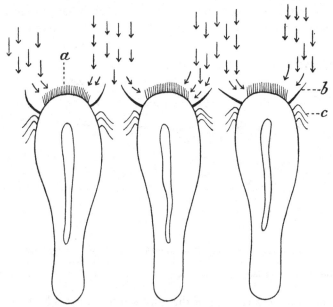

Fig. 99. Transverse section of three gill bars of *Mytilus edulis* shewing *a*, frontal cilia; *b*, latero-frontal cilia; *c*, lateral cilia, also the direction of the currents produced. (After Gray.)

adductor muscles, removing one of the shells and cutting away one side of the mantle (Fig. 98); this exposes the motive power for the main currents, viz. the 'gills.'

On entering the wide mantle cavity the velocity of the inhalent current is reduced and impinging on the curtain *B* heavier particles sink downwards and fall into the grooved edge of the mantle *C*; along this they are driven by cilia working towards the posterior region of the mantle. The main ingoing current with the smaller particles of carmine is drawn over the surface of the gill, and at the same time directed against the face of the filaments. The detailed structure of the gill will be described later; in the meantime it is convenient to follow the visible course of the finer carmine particles. On reaching the surface of the gill the particles partly become embedded in mucus, but all of them are driven downwards to the free distal edge of the gill itself; thence they move forward to the region of the labial palps and so to the mouth.

The water current after leaving the gills eventually reaches the surface of the mantle, whence by the cilia on the latter it is conveyed either directly to the exhalent siphon or by way of the grooved edge, already mentioned, to the same exit.

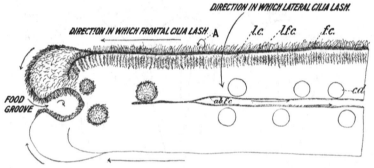

Fig. 100. Lateral view of a gill filament of *Mytilus edulis* (slightly modified from Orton). *l.c.*, lateral cilia; *l.f.c.*, latero-frontal cilia; *f.c.*, frontal cilia; *A*, direction of effective beat of latero-frontal cilia; *ab.f.c.*, abfrontal cilia; *c.d.*, ciliated disc.

The detailed structure of the gills is interesting. There are two gills on each side of the animal and each gill is composed of narrow tubular filaments; each tube is reflexed along its own axis and the two arms are fused together at frequent intervals. Individual filaments are, however, only loosely attached to their neighbours by means of 'ciliated junctions.' The main current of water which is drawn into the mantle cavity is the result of the activity of the *lateral* cilia (see Fig. 99) which are arranged along each side of a filament in three vertical rows. The beat of these cilia is extremely beautiful on account of their very clearly-defined rhythm which passes up one side of the long axis of the filament and down the other. The currents of water set up by these cilia are shewn diagrammatically in Fig. 99.

On meeting the gill surface these sheets of water impinge on the *latero-frontal* cilia whose direction of beat is towards the surface of the filaments. These large *latero-frontal* cilia perform two functions: (*a*) they act as vanes to deflect the wave currents from between the filaments on to the surface of the latter, and (*b*) they appear to keep individual filaments apart, so giving freedom of action to the *lateral* cilia. Having reached the flat surface of the filaments the water current comes under the influence of the

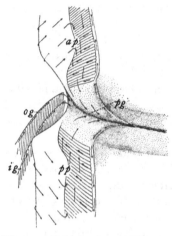

Fig. 101. Labial palps of *Mytilus edulis* as seen from below. *ap*, anterior palp; *ig*, inner demibranch of gill; *og*, outer demibranch of gill; *pg*, proximal oval groove; *pp*, posterior palp. (From Kellogg.)

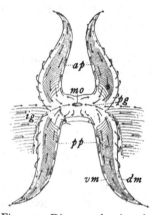

Fig. 102. Diagram shewing the ciliary currents on the labial palps of *Mytilimeria nuttallii*. Mouth and palps seen from below; *ap*, anterior palps; *dm*, dorsal margin of palp; *ig*, inner demibranch of gill; *mo*, mouth; *pg*, proximal oval groove; *pp*, posterior palps; *vm*, ventral margin of palp. (From Kellogg.)

frontal cilia which sweep the included particles down over the flat surface of the gill to its distal edge. During this process the particles become embedded in mucus. On reaching the distal end of the gill, the food stream is diverted forward to the mouth owing to the fact that the frontal cilia situated along the distal edge of the gill are turned so as to direct their effective beat forwards. Some of the food actually travels in the 'food groove,' but by no means all. On reaching the labial palps the food arriving from the gills may either be swept into the mouth, or coming under the influence

G

of the cilia on the dorsal wall of the palp be directed away from the mouth towards the outgoing circulation on the mantle. All recent authors are agreed that the palps possess two ciliary systems which invariably work in opposite directions (see Fig. 101), and the older statements concerning the reversal of individual tracts have never been confirmed.

Condensed into a brief account, the ciliary mechanisms of *Mytilus* appear involved and complicated; as seen in life they present a remarkable and beautiful picture of coordinated movement. If a single mussel is placed in a vessel containing a dense suspension of Indian ink in sea-water, the whole fluid becomes perfectly clear in a surprisingly short time. This fact illustrates the large volume of water which is moved by the ciliated surfaces and is of considerable economic importance. It has been found possible to sterilise mussels by taking advantage of this fact: all traces of infected food particles are rapidly eliminated and the mussels are fit for food.

It has already been pointed out that cilia are sensitive to an accumulation of CO_2 and to a deficiency of oxygen. Excess of CO_2 rapidly depresses ciliary movement, and at the same time cuts down the oxygen requirement of the cells; its effects are very rapidly reversible. On the other hand the effects produced by a scarcity of oxygen are only reversible within certain limits, and the recovery process is always slow. When a mussel is uncovered at low tide, it shuts its shell and thereby ceases to eliminate CO_2; consequently the cilia are brought to rest, and the store of oxygen in the tissues is thereby conserved. As soon as the tide rises and covers the animal with water, the shells open, the cilia very rapidly resume activity, and the animal is thus able to feed at once.

Rotifers.

This group derives its name from the highly developed ciliary organ situated at the anterior end of the body. The very distinct cilia, situated on the disc and on the trochus are the principal organs for collecting food and also for swimming when the animal is not attached to some fixed object by means of its foot. In some cases, as in the Bdelloida, the cilia beat downwards in a plane parallel to the long axis of the body; when the animal is attached by its foot the movement of these cilia produces hollow vortices "like the rings of a skilled cigarette-smoker" (Hartog); when the animal is not attached by its foot, these cilia drive the animal through the water. In other cases, the cilia cause the animal to rotate as well as to progress, and occasionally (*Synchaeta*) a variety of movements can be effected. The trochal cilia when in motion give an illusive impression of the spokes of a wheel owing to the high speed of their effective stroke. The finer cilia of the groove and cingulum play a very minor part in the act of swimming or in the production of the great vortices at the edge of the disc when the animal is stationary; these finer cilia, on the other hand, direct particles towards the mouth and are part of the feeding mechanism. The whole of the gut with the exception of the mastax and cloaca is ciliated.

In some tubicolous forms, cilia play an important rôle in the formation of the tube. In *Melicerta migens* pellets are formed from excess food particles which reach the disc, and being carried thence by ciliated gutters, reach a ciliated glandular cup on the side of the head. Here the particles are rotated with mucus, and so are cemented together to form a pellet which is eventually placed, by muscular contractions, on the edge of the tube.

Protochordata.

The ciliation of *Amphioxus* and of Ascidians has been described by Orton (1913 *a*), and is remarkably similar to that on the gills of Lamellibranch molluscs. In *Amphioxus* the whole of the food is extracted from the water by three separate processes: (i) a current of water is drawn through the pharynx by the activity of the lateral cilia on the sides of the gill-bars; (ii) from the endostyle a stream of mucus is passed on to the internal surface of the pharynx; (iii) the particles of food, being caught in the mucus, are transported by numerous cilia up to the dorsal groove, whence they pass into the digestive tract.

In Ascidians essentially the same system is found. Orton is of opinion that the 'neural gland' of these animals is an organ for secreting mucus, whilst the dorsal tubercle is an organ for passing mucus on to the pharynx. In *Amphioxus* Hatchek's pit and the wheel organ are associated with the collection of food in the buccal cavity.

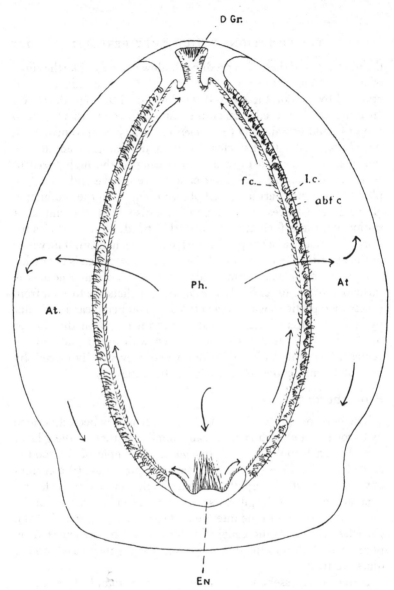

Fig. 103. Diagram of a transverse section of the pharynx and atrium of *Amphioxus* to shew the ciliary mechanisms on the gill, which produce the main current and collect and transport food particles. (This diagram serves equally well for Ascidians.) *Ph.*, pharynx; *At.*, atrium; *En.*, endostyle, the lateral cilia of which pass on mucus and food particles from the median cilia to the gill-bars; *D.Gr.*, dorsal groove of pharynx; *L.c.*, lateral cilia on the gill-bars. These produce the main current which passes across the gill from the pharynx to the atrium in the direction indicated by the large arrows which cross the gill-bars about the middle of the figure; *f.c.*, frontal or pharyngeal cilia. These collect and transport food particles and mucus from the gill surface to the dorsal groove as indicated by the arrows along the inside of the gill-bars. (After Orton.)

Orton (1913 *b*) has pointed out the interesting fact that in Gastropods, Lamellibranchs, Brachiopods, Ascidians and Amphioxus the ciliary mechanisms for food collection are all built upon the same plan.

Vertebrata.

In fishes, reptiles, birds and mammalia there is no longer any trace of locomotory cilia at any time in the life-history. In amphibia on the other hand the organism often moves by cilia both before and after it hatches from the egg. The rotation by cilia of the unhatched larvae which is highly typical of invertebrates was also observed by Balfour in the newt and in the frog. "The

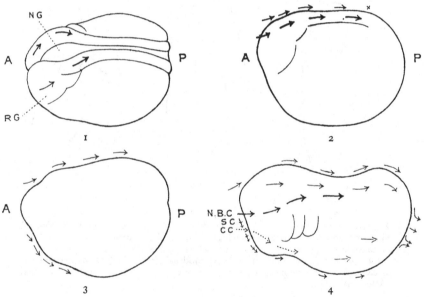

Fig. 104. Ciliary currents on the epidermis of a tadpole of *Rana*. *A*, anterior end; *P*, posterior end. 1. The arrows indicate the extent and position of ciliated area during the upward folding of the neural plate. Surface view. 2. A few hours later than 1; the × indicates the posterior region of the ciliated surface. 3. A second current on the anterior ventral surface has developed. 4. The embryo is now about 3 mm. long, and the whole surface is ciliated. Three well-marked currents are visible: *N.B.C.*, *S.C.*, and *C.C. N.G.*, neural groove; *R.G.*, rudiment of Gasserian and other ganglia; *N.B.C.*, naso-branchial stream; *S.C.*, stomodaeal stream; *C.C.*, current in connection with mucous or cement glands. (After Assheton.)

outer layer of epiblast cells becomes ciliated after the close of the segmentation.... The cilia cause a slow rotatory movement of the embryo within the egg and probably assist in the respiration after it has hatched" (p. 127). At a comparatively early stage, however, the ciliated surface of the ectoderm becomes restricted to particular areas which were investigated by Assheton (1896). The

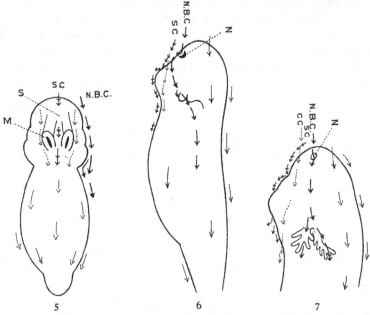

Fig. 105. Ciliary currents on the epidermis of a tadpole (*Rana*). 5. Ventral view of tadpole 3¼ mm. long shewing course of ciliary currents. 6. Lateral view of tadpole 6 mm. long. 7. Side view of tadpole 8 mm. long. *M.*, mucous or cement gland; *S.*, stomodaeum; *S.C.*, stomodaeal stream; *N.B.C.*, naso-branchial stream; *N.*, nasal depression; *C.C.*, current over mucous or cement glands. (After Assheton.)

first of these areas is located in the region of the neural plate and the ciliary current is directed over the dorsal surface towards the tail (Fig. 104).

Just before the neural groove closes a ciliary current becomes obvious on the anterior ventral surface of the embryo (Fig. 104); that is, over the area where the mouth subsequently develops. From this stage onwards the ciliation spreads rapidly, and by the time the embryo is 3 mm. long the whole surface of the embryo is ciliated although the currents produced over different areas vary considerably in intensity. A well-defined and powerful current (Fig. 104) passes over the bases of the branchial arches, whilst less powerful currents pass posteriorly over the dorsal and ventral surfaces. At a later stage (3-4 mm.) the ventral currents become modified by the development of the cement glands (Fig. 105). By the time the tadpole is 6-7 mm. long its cilia are capable of moving the organism over the surface of a flat glass vessel at a rate of 1 mm. in 4-7 seconds. As the embryo increases in size the ciliary currents become less powerful although they still persist in an exhalent flow from the opercular spout and in a surface current over the tail.

In the adult frog cilia are restricted, as in other vertebrata, to the internal surfaces of the body. They are present on the walls of the oesophagus (see Fig. 1), on the bronchi, and in the urogenital system. As one might expect the cilia on the walls of the bronchi beat towards the opening to the oesophagus. In man, similar ciliated tracts exist (Bryant, 1914).

According to Schäfer (1898, p. 201), ciliated epithelium occurs in the following positions in the human body: (i) On the mucous membrane of the air passages and its prolongations, beginning within the nostrils and extending into the nasal duct and lachrymal sac. From the nose it spreads backwards on the upper surface of the soft palate and over the upper or nasal region of the pharynx, thence along the Eustachian tube and the lining membrane of the tympanum. According to Bryant (1914) the velocity of the nasal currents is very low, being approximately 4 mm. per hour, and as in other instances the ciliary expulsion of mucus is sometimes accelerated by muscular activity. (ii) On the walls of the larynx whence it continues throughout the trachea and bronchial tubes in the lungs to their smallest ramifications. The cilia of the bronchial tubes appear to collect mucus into a bolus which is expelled by the muscular movements of coughing. In dogs the currents of the tracheae move with a velocity of 60-290 μ per second (Büttner-Wobst, 1909). (iii) In the Fallopian tubes, where its activity is said to be responsible for the movement of the eggs into the uterus whose walls are also ciliated[1]. It may be noted that the sperma-

---

[1] Lim and Chao (1927) attribute the movement of the rabbit's ova to muscular contractions of the Fallopian tubes and not to the activity of the ciliated epithelium.

tozoa swim against the ciliated current of the Fallopian tubes.
(iv) In the *vasa efferentia* and part of the tube of the epididymis.
(v) On the walls of the ventricles of the brain (except the fifth)
and throughout the central canal of the spinal cord. It would be
interesting to know whether this epithelium is to be regarded as
the homologue of the primitive ectodermal ciliation of vertebrate
embryos (such as is seen in the frog), whose primitive function
may have been locomotory. (vi) In the oesophagus, kidney and
stomach (?) of the embryo.

It is not without interest to note that throughout the animal
kingdom cilia are found in cells derived from each of the three
primary layers, ectoderm, mesoderm and endoderm. Muscular
elements, on the other hand, are normally restricted to cells of
mesodermal origin only.

References

Assheton, R. (1896). Q.J.M.S. 38, *465*.
Bryant, W. S. (1914). Amer. Journ. Physiol. 33, *430*.
Büttner-Wobst, W. (1909). Dissertation, Jena (quoted from Tabulae
    Biologicae, IV).
Carlgren, O. (1905). Biolog. Centralb. 25, *308*.
Elmhirst, R. (1925). Scottish Naturalist, *149*.
Gemmill, J. F. (1915). Proc. Zool. Soc. *1*.
—— (1918). Proc. Zool. Soc. *263*.
—— (1921). Proc. Roy. Phys. Soc. Edin. 20, *222*.
Gislén, T. (1924). *Echinoderm Studies*. Uppsala.
Hartog, M. (1896). Camb. Nat. Hist. 2, *195*.
Kellogg, J. L. (1915). Journ. Morph. 26, *625*.
Lim, R. K. S. and Chao, C. (1927). Chin. Journ. Physiol. 1, *175*.
Orton, J. H. (1912). Journ. Mar. Biol. Assoc. 9, *444*.
—— (1913). Journ. Mar. Biol. Assoc. 10, *19*.
—— (1913 b). Journ. Mar. Biol. Assoc. 10, *283*.
—— (1922). Nature, 110, *178*.
Parker, G. H. (1896). Bull. Mus. Comp. Zool. Harvard, 29, *107*.
—— (1905). Amer. Journ. Physiol. 13, *1*.
Rees, C. W. (1922). Univ. Calif. Publ. Zool. 20, *333*.
Schäfer, E. A. (1898). *Quain's Elements of Anatomy*. Vol. 1, Part 2
    London.
Taylor, C. V. (1920). Univ. Calif. Publ. Zool. 19, *404*.
Wallengren, H. (1905). Lunds Univ. Årsskr. N.F. Afd. 2, Bd. 1.
Widmark, E. M. P. (1913). Zeit. f. Allg. Physiol. 15, *33*.
Yonge, C. M. (1926). Journ. Mar. Biol. Assoc. 14, *295*.

BIBLIOGRAPHY

Alexeieff, A. (1911). Sur les Cercomonadines intestinales de *Calliphora erythrocephala* Mg. et de *Lucilia sp.* C. R. Soc. Biol. 71, *379.*

Allen, W. R. (1914). The Food and Feeding Habits of Freshwater Mussels. Biol. Bull. 27, *127.*

—— (1921). Studies of the Biology of Freshwater Mussels. Biol. Bull. 40, *210.*

Alverdnes, F. (1922). Untersuchungen über begeisselte und beflimmerte Organismen. Arch. Entw. Mech. 52, *281.*

v. Angerer (1919). Über die Arbeitsleistung eigenbeweglicher Bakterien. Archiv f. Hygiene, 88, *139.*

Assheton, R. (1896). Notes on the Ciliation of the Ectoderm of the Amphibian Embryo. Quart. Journ. Micr. Sci. 38, *465.*

Ballowitz, E. (1889). Fibrilläre Struktur und Contraktilität. Pflüger's Archiv, 46, *433.*

Bayliss, W. (1918). *Principles of Gen. Physiol.* London.

Becker, O. (1857). Molescholt's Untersuch. 2, *71*—(from Pütter).

Benda, C. (1901). Über neue Darstellungsmethoden der Zentralkörperchen und die Verwandtschaft der Basalkörper der Cilien mit Zentralkörperchen. Verh. Physiol. Ges. Berlin.

Bhatia, D. (1926). The Structure of the Latero-frontal Cells of the Gills of *Mytilus edulis.* Quart. Journ. Micr. Sci. 70, *681.*

Bidder, G. P. (1923). The Relation of the Form of a Sponge to its Currents. Quart. Journ. Micr. Sci. 67, *293.*

Bowditch, H. P. (1876). Force of Ciliary Motion. Boston Med. and Surg. Journ. 15, *157.*

v. Brücke, E. T. (1916). Versuche an ausgeschnittenen und nach einer Drehung um 180° reimplantierten Flimmerschleimhaut-Stücken. Pflüger's Archiv, 166, *45.*

Bryant, W. S. (1914). An Experiment to prove that the Cilia of the Human Nose waft toward the Anterior Nares. Amer. Journ. Physiol. 33, *430.*

Bütschli, O. (1889). Bronn's *Klassen und Ordnungen. Protozoa, 3.* Leipzig.

—— (1889). Bronn's Tierreich. Bd. 1, *2035.*

Büttner-Wobst, W. (1909). Über die Flimmerbewegung in Trachea und Bronchien des lebenden Säugetieres. Dissert. Jena.

Calkins, G. N. (1926). *Biology of the Protozoa.* London.

Calliburces, P. (1858). Recherches expérimentales sur l'Influence exercée par la Chaleur sur les Manifestations de la Contractilité des Organes. Comptes Rendus, 47, *638.*

Carlgren, O. (1905). Über die Bedeutung der Flimmerbewegung für den Nahrungstransport bei den Actiniarien und Madreporarien. Biolog. Centralb. 25, *308*.

Carter, G. S. (1924). On the Structure and Movements of the Latero-frontal Cilia of the Gills of *Mytilus*. Proc. Roy. Soc. 96 B, *115*.

—— (1926). On the Nervous Control of the Velar Cilia of the Nudibranch Veliger. Brit. Journ. Exp. Biol. 4, *1*.

Chambers, R. and **Dawson, J. A.** (1925). The Structure of the Undulating Membrane in the Ciliate *Blepharisma*. Biol. Bull. 48, *240*.

Cole, A. E. (1921). Oxygen supply of certain Animals living in Water containing no dissolved Oxygen. Journ. Exp. Zool. 33, *293*.

Copeland, M. (1919). Locomotion in two species of the Gastropod genus *Alectrion* with Observations on the Behavior of Pedal Cilia. Biol. Bull. 37, *126*.

—— (1922). Ciliary and Muscular Locomotion in the Gastropod genus *Polinices*. Biol. Bull. 42, *132*.

Dale, D. and **Thacker, C. R. A.** (1914). Hydrogen Ion Concentrations limiting Automaticity in different Regions of the Frog's Heart. Journ. of Physiol. 47, *493*.

Dellinger, O. P. (1909). The Cilium as a key to the Structure of Contractile Protoplasm. Journ. Morph. 20, *171*.

Duboscq, O. (1907). Sur la Motilité des Filaments Axiles dans les Spermatozoïdes Géants de la Paludine. Comptes Rendus Assoc. Anat. 9, *130*.

Elmhirst, R. (1925). Associations between the Amphipod Genus *Metopa* and Coelenterates. II. The Feeding Habits of the Sea-Anemone, *Actinoloba*. Scottish Naturalist, *149*.

Embden, G. (1923). Über die Bedeutung von Ionen für den Chemismus der Muskelkontraktion und den Ablauf fermentativer Reaktionen. Naturwiss. 11, *985*.

Engelmann, T. W. (1868). Ueber die Flimmerbewegung. Jena. Zeit. 4, *321*.

—— (1875). Contractilität und Doppelbrechung. Pflüger's Archiv, 11, *432*.

—— (1877). Flimmerruhr und Flimmermühle. Zwei Apparate zum Registriren der Flimmerbewegung. Pflüger's Archiv, 15, *493*.

—— (1898). *Dictionnaire de Physiologie*. Paris.

Erhard, H. (1910). Studien über Flimmerzellen. Archiv f. Zellforsch. 4, *309*.

—— (1922). Abderhalden's *Handbuch der Biol. Arbeitsmeth.* Abt. v. 2, 3, *213*.

Evans, C. L. (1912). The Gaseous Metabolism of the Heart and Lungs. Journ. of Physiol. 45, *213*.

Fedele, M. (1923). Le attività dinamiche ed i rapporti nervosi nella vita dei Dolioli. Pubblicazioni della Stazione Zoologica di Napoli, 4, *129*.

—— (1924). Le prove sperimentali di una regolazione nervosa del movimento ciliare. Pubblic. d. Staz. Zool. d. Napoli, 5, *275*.

Frédéricq, H. (1913). Recherches expérimentales sur la Physiologie Cardiaque d'*Octopus vulgaris*. Bull. Acad. Roy. Belgique, *758*.

Fröhlich, F. W. (1905). Ueber die scheinbare Steigerung der Leistungsfähigkeit des quergestreiften Muskels im Beginn der Ermüdung ("Muskeltreppe"), der Kohlensäurewirkung und der Wirkung anderer Narkotika (Aether, Alkohol). Zeit. f. Allg. Physiol. 5, *288*.

Fuhrmann, F. (1910). Die Geisseln von *Spirillum volutans*. Centralb. f. Bakter. II. Abt. 25, *129*.

Gemmill, J. F. (1915). On the Ciliation of Asterids, and on the Question of Ciliary Nutrition in Certain Species. Proc. Zool. Soc. *1*.

—— (1918). Ciliary Action in the Internal Cavities of the Ctenophore *Pleurobrachia pileus* Fabr. Proc. Zool. Soc. *263*.

—— (1921). Notes on Food-Capture and Ciliation in the Ephyrae of *Aurelia*. Proc. Roy. Phys. Soc. Edinb. 20, *222*.

Gislén, T. (1924). *Echinoderm Studies*. Zoologiska Bidrag från Uppsala, 9. Uppsala.

Goldschmidt, R. (1907). Lebensgeschichte der Mastigamöben *Mastigella vitrea* n. sp. u. *Mastigina setosa* n. sp. Archiv f. Protistenk. Suppl. 1, *83*.

Göthlin, G. F. (1920). Inhibition of Ciliary Movement in *Beroë cucumis*. Journ. Exp. Zool. 31, *403*.

Grave, C. and **Schmidt, F. O.** (1925). A Mechanism for the Co-ordination and Regulation of Ciliary Movement, etc. Journ. Morph. 40, *479*.

Gray, J. (1920). The Effects of Ions upon Ciliary Movement. Quart. Journ. Micr. Sci. 64, *345*.

—— (1922). The Mechanism of Ciliary Movement. Proc. Roy. Soc. 93 B, *104*.

—— (1922). The Mechanism of Ciliary Movement. II. The Effect of Ions on the Cell Membrane. Proc. Roy. Soc. 93 B, *122*.

—— (1923). The Mechanism of Ciliary Movement. III. The Effect of Temperature. Proc. Roy. Soc. 95 B, *6*.

—— (1924). The Mechanism of Ciliary Movement. IV. The Relation of Ciliary Activity to Oxygen Consumption. Proc. Roy. Soc. 96 B, *95*.

—— (1926). The Mechanism of Ciliary Movement. V. The Effect of Ions on the Duration of Beat. Proc. Roy. Soc. 99 B, *398*.

—— (1926). The Properties of an Intercellular Matrix and its Relation to Electrolytes. Brit. Journ. Exp. Biol. 3, *167*.

Gurwitsch, A. (1901). Studien über Flimmerzellen. Archiv f. Mikr. Anat. 57, *184*.

—— (1904). *Morphologie und Biologie der Zelle*. Jena, 1904.

Hartog, M. (1896). Camb. Nat. Hist. 2, *195*.

Hartree, W. and **Hill, A. V.** (1922). The Heat-production and the Mechanism of the Veratrine Contraction. Journ. of Physiol. 56, *294*.

Haywood, C. (1925). The Relative Importance of *p*H and Carbon Dioxide Tension in determining the Cessation of Ciliary Movement in Acidified Water. Journ. Gen. Physiol. 7, *693*.

Heidenhain, M. (1911). *Plasma und Zelle*, 1, 2. Jena.

Hensen, V. (1881). Hermann's *Handbuch der Physiol*. Bd. 6, *89*.

Hertwig, R. (1877). Studien über Rhizopoden. Jena. Zeit. 11, *324*.

Hesse, R. (1900). Untersuchungen über die Organe der Lichtempfindung bei niederen Thieren. Zeit. f. wiss. Zool. 68, *379*.

Hill, A. V. and **Hartree, W.** (1920). The Thermo-elastic Properties of Muscle. Phil. Trans. Roy. Soc. 210 B, *153*.

—— —— (1920). The Four Phases of Heat-production of Muscle. Journ. of Physiol. 54, *84*.

Höber, R. (1914). *Physik. Chem. der Zelle und Gewebe*, *508, 532*.

Hogben, L. T. (1925). Studies on the Comparative Physiology of Contractile Tissues. I. The Action of Electrolytes on Invertebrate Muscle. Quart. Journ. Exp. Physiol. 15, *263*.

Iijuma, I. (1884). Untersuchungen über den Bau und die Entwicklungsgeschichte der Süsswasser-Dendrocoelen (Tricladen). Zeit. f. wiss. Zool. 40, *359*.

Inchley, O. (1921). A simple Apparatus to demonstrate Activity of Cilia. Proc. Physiol. Soc. cxxvii.

James-Clarke, H. (1868). On the *Spongiae ciliatae* as *Infusoria flagellata*; or Observations on the Structure, Animality, and Relationship of *Leucosolenia botryoides*, Bowerbank. Annals of Nat. Hist. 4th series, 1, *133, 188, 250*.

Jennings, H. S. (1915). *Behaviour of the Lower Organisms*. New York.

Jensen, P. (1893). Die absolute Kraft einer Flimmerzelle. Pflüger's Archiv, 54, *537*.

Jordan, A. E. and **Helvestine, F.** (Jnr) (1922). Ciliogenesis in the Epididymis of the white Rat. Anat. Record, 25, *7*.

Joseph, H. (1902). Beiträge zur Flimmerzellen- und Centrosomenfrage. Arb. aus d. Zool. Inst. d. Univ. Wien, 14, *1*.

Kaiser, J. E. (1893). Die Acanthocephalen und ihre Entwickelung. Biblioth. Zool. 2 (vii).

Kellogg, J. L. (1915). Ciliary Mechanisms of Lamellibranchs. Journ. Morph. 26, *625*.

Kindred, J. E. (1926). Cell Division and Ciliogenesis in the Ciliated Epithelium of the Pharynx and Esophagus of the Tadpole of the green Frog, *Rana clamitans*. Journ. of Morph. and Physiol. 43, *267*.

Kolm, R. and **Pick, E. P.** (1920). Über die Bedeutung des Kaliums für die Selbststeuerung des Herzens. Pflüger's Archiv, 185, *235*.

Kraft, H. (1890). Zur Physiologie des Flimmerepithels bei Wirbelthieren. Pflüger's Archiv, 47, *196*.

Krijgsman, B. J. (1925). Beiträge zum Problem der Geisselbewegung. Archiv f. Protistenk. 52, *478*.

Kugelmass, I. N. and **Shohl, A. T.** (1923). Equilibria involving Calcium, Hydrogen, Carbonate, Bicarbonate, Primary, Secondary and Tertiary Phosphate Ions. Proc. Soc. Exp. Biol. and Med. 21, *6*.

Lapage, G. (1925). Notes on the Choanoflagellate, *Codosiga botrytis*, Ehrbg. Quart. Journ. Micr. Sci. 69, *471*.

Lillie, R. S. (1906). The Relation between Contractility and Coagulation of the Colloids in the Ctenophore Swimming-plate. Amer. Journ. Physiol. 16, *117*.

—— (1906). The Relation of Ions to Contractile Processes. I. The Action of Salt Solutions on the Ciliated Epithelium of *Mytilus edulis*. Amer. Journ. Physiol. 17, *89*.

Lim, R. K. S. and **Chao, C.** (1927). On the Mechanism of the Transplantation of Ova. I. Rabbit Uterus. Chin. Journ. Physiol. 1, *175*.

Mackinnon, D. L. and **Vlès, F.** (1908). On the Optical Properties of Contractile Organs. Journ. Roy. Micr. Soc. *553*.

Maier, H. N. (1903). Über den feineren Bau der Wimperapparate der Infusorien. Archiv f. Protistenk. 2, *73*.

Mast, S. O. and **Nadler, J. E.** (1926). Reversal of Ciliary Action in *Paramecium caudatum*. Journ. Morph. 43, *105*.

Maxwell, S. S. (1905). The Effect of Salt-solutions on Ciliary Activity. Amer. Journ. Physiol. 13, *154*.

Mayer, A. G. (1906). Rhythmical Pulsation in Scyphomedusae. Publ. Carnegie Instit. No. 47.

—— (1911). The Converse Relation between Ciliary and Neuro-Muscular Movements. Publ. Carnegie Instit. No. 132.

McDonald, J. F., Leisure, C. E. and **Lenneman, E. E.** (1927). Neural and Chemical Control of Ciliated Epithelium. Proc. Soc. Exp. Biol. and Med. 24, *968*.

Merton, H. (1923). Studien über Flimmerbewegung. Pflüger's Archiv, 198, *1*.

—— (1923 *a*). "Willkürliche" Flimmerbewegung bei Metazoen. Biol. Zentralb. 43, *157*.

—— (1924). Lebenduntersuchungen an den Zwitterdrüsen der Lungenschnecken. Zeit. f. wiss. Biol. (Zellen und Gewebelehre), 1, *671*.

Metzner, P. (1920). Zur Mechanik der Geisselbewegung. Biol. Zentralb. 40, *49*.

Meves, F. (1899). Ueber Struktur und Histogenese der Samenfäden des Meerschweinchens. Archiv f. Mikr. Anat. 54, *381*.

Miklucho-Maclay, N. (1868). Beiträge zur Kenntniss der Spongien I. Jena. Zeit. 4, *221*.

Minchin, E. A. (1912). *Introduction to the Study of the Protozoa.* London.

Mines, G. R. (1913). On Functional Analysis by the Action of Electrolytes. Journ. of Physiol. 46, *188*.

Minot, C. S. (1877). Studien an Turbellarien. Beiträge zur Kenntniss der Plathelminthen. Arb. Zool.-Zootom. Inst. Würzburg, 3, *405*.

de Morgan, W. (1926). Further Observations on Marine Ciliates Living in the Laboratory Tanks at Plymouth. Journ. Mar. Biol. Assoc. 14, *23*.

Nelson, T. C. (1923). The Mechanism of Feeding in the Oyster. Proc. Soc. Exp. Biol. and Med. 21, *166*.

Orton, J. H. (1912). The Mode of Feeding of *Crepidula*, with an Account of the Current-producing Mechanism in the Mantle Cavity, and some remarks on the Mode of Feeding in Gastropods and Lamellibranchs. Journ. Mar. Biol. Assoc. 9, *444*.

—— (1913). The Ciliary Mechanisms on the Gill and the Mode of Feeding in Amphioxus, Ascidians, and *Solenomya togata*. Journ. Mar. Biol. Assoc. 10, *19*.

—— (1914). On Ciliary Mechanisms in Brachiopods and some Polychaetes, with a Comparison of the Ciliary Mechanisms on the Gills of Molluscs, Protochordata, Brachiopods, and Cryptocephalous Polychaetes, and an Account of the Endostyle of Crepidula and its Allies. Journ. Mar. Biol. Assoc. 10, *283*.

—— (1922). The Mode of Feeding of the Jelly-fish, *Aurelia aurita*, on the Smaller Organisms in the Plankton. Nature, 110, *178*.

Ostwald, W. (1900). Periodische Erscheinungen bei der Auflösung des Chroms in Säuren. Zeit. f. Physikal. Chem. 35, *33*.

Parker, G. H. (1896). The Reactions of *Metridium* to Food and other Substances. Bull. Mus. Comp. Zool. Harvard, 29, *107*.

—— (1905). The Reversal of Ciliary Movement in Metazoans. Amer. Journ. Physiol. 13, *1*.

—— (1905). The Movements of the Swimming-plates in Ctenophores, with reference to the theories of Ciliary Metachronism. Journ. Exp. Zool. 2, *407*.

—— (1918). *The Elementary Nervous System.* Philadelphia.

Perrin, W. S. (1906). Researches upon the Life-history of *Trypanosoma balbianii* (Certes). Archiv f. Protistenk. 7, *131*.

Peter, K. (1899). Das Centrum für die Flimmer- und Geisselbewegung. Anat. Anz. 15, *271*.

Plate, L. (1888). Studien über Protozoen. Zool. Jahrb. Anat. 3, *135*.

Prénant, A. (1913). Les Appareils Ciliés et leurs Dérivés. Journ. de l'Anat. et Physiol. 49, *150*, *545*.

Prowazek, S. (1902). Protozoenstudien III. *Euplotes harpa*. Arb. aus d. Zool. Inst. d. Univ. Wien, 14, *81*.

Pütter, A. (1903). Die Flimmerbewegung. Ergebn. der Physiol. 2, 11, *1*.

—— (1905). Die Atmung der Protozoen. Zeit. f. Allg. Physiol. 5, *566*.

Rees, C. W. (1922). The Neuromotor Apparatus of *Paramecium*. Univ. Calif. Publ. Zool. 20, *333*.

Reichert, K. (1909). Ueber die Sichtbarmachung der Geisseln und die Geisselbewegung der Bakterien. Centralb. f. Bakter. I. Abt. 51, *14*.

Saguchi, S. (1917). Studies on Ciliated Cells. Journ. Morph. 29, *217*.

Sakai, T. (1914). Über die Wirkung einiger Anionen auf den isolierten Froschventrikel. Zeit. f. Biol. 64, *1*.

Saunders, J. T. (1925). The Trichocysts of *Paramecium*. Proc. Camb. Philos. Soc. Biol. Sci. 1, *249*.

Schäfer, E. A. (1891). On the Structure of Amoeboid Protoplasm, with a Comparison between the Nature of the Contractile Process in Amoeboid Cells and in Muscular Tissue, and a Suggestion re-garding the Mechanism of Ciliary Action. Proc. Roy. Soc. 49, *193*.

—— (1898). *Quain's Elements of Anatomy*, Vol. 1, Pt 2. London.

—— (1904). Theories of Ciliary Movement. Anat. Anz. 24, *497*.

—— (1905). Models to illustrate Ciliary Action. Anat. Anz. 26, *517*.

Segerdahl, E. (1922). Investigations on (1) the Effect of a Direct Electric Current, (2) of some Chemicals, on the Ciliary Motion of the Anodonta Gill. Skand. Arch. Physiol. 42, *62*.

Statkewitsch, P. (1905). Galvanotropismus und Galvanotaxis der Ciliata II. Zeit. Allg. Phys. 5, *511*.

Taylor, C. V. (1920). Demonstration of the Function of the Neuro-motor Apparatus in *Euplotes* by the Method of Micro-dissection. Univ. Calif. Publ. Zool. 19, *403*.

Torrey, H. B. (1904). On the Habits and Reactions of *Sagartia davisi*. Biol. Bull. 6, *203*.

Trigt, H. van (1919). *A Contribution to the Physiology of the Fresh-water Sponges*. Leiden.

Valentin, G. (1842). Wagner's *Handwörterb. der Physiol.* 1, *484*.

Verworn, M. (1890). Studien zur Physiologie der Flimmerbewegung. Pflüger's Archiv, 48, *149*.

Vignon, P. (1901). Recherches de Cytologie Générale sur les Épithéliums. Archives de Zool. Exp. et Gén., Sér. 3, t. 9, *371*.

Wallengren, H. (1905). Zur Biologie der Muscheln. I. Die Wasser-strömungen. Lunds Universitets Årsskrift, N.F. Afd. 2, Bd. 1, No. 2, *1*.

Wenyon, C. M. (1926). *Protozoology*, vol. 1. London.

Widmark, E. M. P. (1913). Über die Wasserströmungen in dem Gastrovaskularapparat von *Aurelia aurita* L. Zeit. f. Allg. Physiol. *15, 33*.

Williams, L. W. (1907). The Structure of Cilia, especially in Gastropods. Amer. Naturalist, 41, *545*.

Woerdeman, M. W. (1925). Entwicklungsmechanische Untersuchungen über die Wimperbewegung des Ectoderms von Amphibienlarven. Zeit. f. Wiss. Biol. Abt. D, 106, *41*.

Wyman, J. (1925). Neuroid Transmission in Ciliated Epithelium. Journ. Gen. Physiol. 7, *545*.

Yonge, C. M. (1925). The Hydrogen Ion Concentration in the Gut of certain Lamellibranchs and Gastropods. Journ. Mar. Biol. Assoc. 13, *938*.

—— (1926). Structure and Physiology of the Organs of Feeding and Digestion in *Ostrea edulis*. Journ. Mar. Biol. Assoc. 14, *295*.

SUBJECT INDEX

Printed in the United States
By Bookmasters